PROGRESS IN CLINICAL AND BIOLOGICAL RESEARCH

RECENT TITLES

Vol 314: **Inherited and Environmentally Induced Retinal Degenerations,** Matthew M. LaVail, Robert E. Anderson, Joe G. Hollyfield, *Editors*

Vol 315: **Muscle Energetics,** Richard J. Paul, Gijs Elzinga, Kazuhiro Yamada, *Editors*

Vol 316: **Hemoglobin Switching,** George Stamatoyannopoulos, Arthur W. Nienhuis, *Editors*. Published in two volumes: Part A: *Transcriptional Regulation*. Part B: *Cellular and Molecular Mechanisms*.

Vol 317: **Alzheimer's Disease and Related Disorders,** Khalid Iqbal, Henryk M. Wisniewski, Bengt Winblad, *Editors*

Vol 318: **Mechanisms of Chromosome Distribution and Aneuploidy,** Michael A. Resnick, Baldev K. Vig, *Editors*

Vol 319: **The Red Cell: Seventh Ann Arbor Conference,** George J. Brewer, *Editor*

Vol 320: **Menopause: Evaluation, Treatment, and Health Concerns,** Charles B. Hammond, Florence P. Haseltine, Isaac Schiff, *Editors*

Vol 321: **Fatty Acid Oxidation: Clinical, Biochemical, and Molecular Aspects,** Kay Tanaka, Paul M. Coates, *Editors*

Vol 322: **Molecular Endocrinology and Steroid Hormone Action,** Gordon H. Sato, James L. Stevens, *Editors*

Vol 323: **Current Concepts in Endometriosis,** Dev R. Chadha, Veasy C. Buttram, Jr., *Editors*

Vol 324: **Recent Advances in Hemophilia Care,** Carol K. Kasper, *Editor*

Vol 325: **Alcohol, Immunomodulation, and AIDS,** Daniela Seminara, Ronald Ross Watson, Albert Pawlowski, *Editors*

Vol 326: **Nutrition and Aging,** Derek M. Prinsley, Harold H. Sandstead, *Editors*

Vol 327: **Frontiers in Smooth Muscle Research,** Nicholas Sperelakis, Jackie D. Wood, *Editors*

Vol 328: **The International Narcotics Research Conference (INRC) '89,** Rémi Quirion, Khem Jhamandas, Christina Gianoulakis, *Editors*

Vol 329: **Multipoint Mapping and Linkage Based Upon Affected Pedigree Members: Genetic Analysis Workshop 6,** Robert C. Elston, M. Anne Spence, Susan E. Hodge, Jean W. MacCluer, *Editors*

Vol 331: **Mouse Liver Carcinogenesis: Mechanisms and Species Comparisons,** Donald E. Stevenson, R. Michael McClain, James A. Popp, Thomas J. Slaga, Jerrold M. Ward, Henry C. Pitot, *Editors*

Vol 332: **Molecular and Cellular Regulation of Calcium and Phosphate Metabolism,** Meinrad Peterlik, Felix Bronner, *Editors*

Vol 333: **Bone Marrow Purging and Processing,** Samuel Gross, Adrian P. Gee, Diana A. Worthington-White, *Editors*

Vol 334: **Potassium Channels: Basic Function and Therapeutic Aspects,** Thomas J. Colatsky, *Editor*

Vol 335: **Evolution of Subterranean Mammals at the Organismal and Molecular Levels,** Eviatar Nevo, Osvaldo A. Reig, *Editors*

Vol 336: **Dynamic Interactions of Myelin Proteins,** George A. Hashim, Mario Moscarello, *Editors*

Vol 337: **Apheresis,** Gail Rock, *Editor*

Vol 338: **Hematopoietic Growth Factors in Transfusion Medicine,** Jerry Spivak, William Drohan, Douglas Dooley, *Editors*

Vol 339: **Advances in Cancer Control: Screening and Prevention Research,** Paul F. Engstrom, Barbara Rimer, Lee E. Mortenson, *Editors*

Vol 340: **Mutation and the Environment,** Mortimer L. Mendelsohn, Richard J. Albertini, *Editors*. Published in five volumes: Part A: *Basic Mechanisms*. Part B: *Metabolism, Testing Methods, and Chromosomes*. Part C: *Somatic and Heritable Mutation, Adduction, and Epidemiology*. Part D: *Carcinogenesis*. Part E: *Environmental Genotoxicity, Risk, and Modulation*.

Vol 341: **Chronobiology: Its Role in Clinical Medicine, General Biology, and Agriculture,** Dora K. Hayes, John E. Pauly, Russel J. Reiter, *Editors*. Published in two volumes.

Vol 342: **Progress in Comparative Endocrinology,** August Epple, Colin G. Scanes, Milton H. Stetson, *Editors*

Vol 343: **Horizons in Membrane Biotechnology,** Claude Nicolau, Dennis Chapman, *Editors*

Vol 344: **Isozymes: Structure, Function, and Use in Biology and Medicine,** Zen-Ichi Ogita, Clement L. Markert, *Editors*

Vol 345: **Sleep and Respiration,** Faiq G. Issa, Paul M. Suratt, John E. Remmers, *Editors*

Vol 346: **Recent Progress in Research on Nutrition and Cancer,** Curtis J. Mettlin, Kunio Aoki, *Editors*

Vol 347: **Mutagens and Carcinogens in the Diet,** Michael W. Pariza, Hans-Ulrich Aeschbacher, James S. Felton, Shigeaki Sato, *Editors*

Vol 348: **EORTC Genitourinary Group Monograph 9: Basic Research and Treatment of Renal Cell Carcinoma Metastasis,** C.G. Bollack, D. Jacqmin, *Editors*

Vol 349: **Cytokines and Lipocortins in Inflammation and Differentiation,** Marialuisa Melli, Luca Parente, *Editors*

Vol 350: **Uro-Oncology: Current Status and Future Trends,** H.G.W. Frohmüller, Manfred Wirth, *Editors*

Vol 351: **Taurine: Functional Neurochemistry, Physiology, and Cardiology,** Herminia Pasantes-Morales, David L. Martin, William Shain, Rafael Martín del Río, *Editors*

Vol 352: **The Biology of Hematopoiesis,** Nicholas Dainiak, Eugene P. Cronkite, Ronald McCaffrey, Richard K. Shadduck, *Editors*

Vol 353: **Neoadjuvant Chemotherapy in Invasive Bladder Cancer,** Ted A.W. Splinter, Howard I. Scher, *Editors*

Vol 354: **Effects of Therapy on Biology and Kinetics of the Residual Tumor,** Joseph Ragaz, Linda Simpson-Herren, Marc Lippman, Bernard Fisher, *Editors*. Published in two volumes: Part A: *Pre-Clinical Aspects*. Part B: *Clinical Aspects*.

Vol 355: **Radiolabelled Cellular Blood Elements,** Helmut Sinzinger, Mathew L. Thakur, *Editors*

Vol 356: **Molecular Biology and Differentiation of Megakaryocytes,** Janine Breton-Gorius, Jack Levin, Alan T. Nurden, Neil Williams, *Editors*

Vol 357: **EORTC Genitourinary Group Monograph 7: Prostate Cancer and Testicular Cancer,** Donald W.W. Newling, William G. Jones, *Editors*

Vol 358: **Treatment of Post Surgical Adhesions,** Gere S. diZerega, L. Russell Malinak, Michael P. Diamond, Cary B. Linsky, *Editors*

Vol 359: **EORTC Genitourinary Group Monograph 8: Treatment of Prostatic Cancer—Facts and Controversies,** Fritz H. Schröder, *Editor*

Vol 360: **Molecular Genetics of Chromosome 21 and Down Syndrome,** David Patterson, Charles J. Epstein, *Editors*

Vol 361: **Current and Future Trends in Anticonvulsant, Anxiety, and Stroke Therapy,** Brian S. Meldrum, Michael Williams, *Editors*

Vol 362: **The Molecular Biology of the Retina: Basic and Clinically Relevant Studies,** Debora B. Farber, Gerald J. Chader, *Editors*

Vol 363: **Information Processing in Medical Imaging,** Douglas A. Ortendahl, Jorge Llacer, *Editors*

Please contact publisher for information about previous titles in this series.

THE MOLECULAR BIOLOGY OF THE RETINA

Basic and Clinically Relevant Studies

THE MOLECULAR BIOLOGY OF THE RETINA

Basic and Clinically Relevant Studies

Proceedings of the Eighth Annual International Society of Eye Research Meeting, Held in San Francisco, California, September 1–5, 1988

Editors

Debora B. Farber
Jules Stein Eye Institute
Department of Ophthalmology
UCLA School of Medicine
Los Angeles, California

Gerald J. Chader
Laboratory of Retinal Cell
and Molecular Biology
National Eye Institute
National Institutes of Health
Bethesda, Maryland

WILEY-LISS

A JOHN WILEY & SONS, INC., PUBLICATION
NEW YORK • CHICHESTER • BRISBANE • TORONTO • SINGAPORE

Address all Inquiries to the Publisher
Wiley-Liss, Inc., 41 East 11th Street, New York, NY 10003

Printed in United States of America

The publication of this volume was facilitated by the authors and editors who submitted the text in a form suitable for direct reproduction without subsequent editing or proofreading by the publisher.

Library of Congress Cataloging-in-Publication Data

International Society for Eye Research. Meeting (8th : 1988 : San Francisco, Calif.)
The molecular biology of the retina : basic and clinically relevant studies : proceedings of the Eighth Annual International Society of Eye Research Meeting held in San Francisco, September 1-5, 1988 / editors, Debora B. Farber, Gerald J. Chader.
p. cm. -- (Progress in clinical and biological research ; v. 362)
Includes bibliographical references and index.
ISBN 0-471-56069-3
1. Retina--Diseases--Molecular aspects--Congresses. 2. Retina--Congresses. 3. Molecular biology--Congresses. I. Farber, Debora. II. Chader, Gerald. III. Title. IV. Series.
[DNLM: 1. Genetics, Biochemical--congresses. 2. Retina--congresses. 3. Retinal Diseases--genetics--congresses. W1 PR668E v.362 / WW 270 I625m 1988]
RE551.I58 1988
617.7'3--dc20
DNLM/DLC
for Library of Congress 90-13055
CIP

Contents

Contributors

T. Abe, Molecular Biology Section, Laboratory of Retinal Cell and Molecular Biology, National Eye Institute, National Institutes of Health, Bethesda, MD 20892 **[163]**

Wolfgang Baehr, Department of Ophthalmology, Cullen Eye Institute, Baylor College of Medicine, Houston, TX 77030 **[33, 115]**

Diane E. Borst, Laboratory of Retinal Cell and Molecular Biology, National Eye Institute, National Institutes of Health, Bethesda, MD 20892 **[139]**

Cathy Bowes, Jules Stein Eye Institute and Department of Ophthalmology, UCLA School of Medicine, Los Angeles, CA 90024-7008 **[67]**

Gerald J. Chader, Laboratory of Retinal Cell and Molecular Biology, National Eye Institute, National Institutes of Health, Bethesda, MD 20892 **[139]**

Carolyn Chambers, National Eye Institute, National Institutes of Health, Bethesda, MD 20892 **[191]**

Michael Danciger, Jules Stein Eye Institute and Department of Ophthalmology, UCLA School of Medicine, Los Angeles, CA 90024-7008, and Department of Biology, Loyola Marymount University, Los Angeles, CA 90045 **[67]**

Larry A. Donoso, Wills Eye Hospital, Philadelphia, PA 19107 **[221]**

Debora B. Farber, Jules Stein Eye Institute and Department of Ophthalmology, UCLA School of Medicine, Los Angeles, CA 90024-7008 **[67]**

Li Geng, Department of Ophthalmology, Cullen Eye Institute, Baylor College of Medicine, Houston, TX 77030 **[115]**

Yoshihiro Hotta, National Eye Institute, National Institutes of Health, Bethesda, MD 20892; present address: Department of Ophthalmology, Juntendo University School of Medicine, Bunkyo-ku, Tokyo, 113, Japan **[191]**

Shi Huang, Department of Pathology, University of California, San Diego School of Medicine, San Diego, CA 92307 **[221]**

George Inana, National Eye Institute, National Institutes of Health, Bethesda, MD 20892; present address: Department of Ophthalmology, Bascomb Palmer Eye Institute, University of Miami School of Medicine, Miami, FL 33136 **[191]**

Nancy G. Kennaway, Department of Medical Genetics, Oregon Health Sciences University, Portland, OR 97201 **[191]**

Eva Y.–H.P. Lee, Department of Pathology, University of California, San Diego School of Medicine, San Diego, CA 92307 **[221]**

Wen-Hwa Lee, Department of Pathology, University of California, San Diego School of Medicine, San Diego, CA 92307 **[221]**

The numbers in brackets are the opening page numbers of the contributors' articles.

Gregory I.H. Liou, Department of Ophthalmology, Cullen Eye Institute, Baylor College of Medicine, Houston, TX 77030 [**115**]

Akira Nakajima, Department of Ophthalmology, Juntendo University School of Medicine, Tokyo, Japan [**191**]

John M. Nickerson, Laboratory of Retinal Cell and Molecular Biology, National Eye Institute, National Institutes of Health, Bethesda, MD 20892 [**139**]

William L. Pak, Department of Biological Sciences, Purdue University, West Lafayette, IN 47907 [**1**]

Steven J. Pittler, Department of Ophthalmology, Cullen Eye Institute, Baylor College of Medicine, Houston, TX 77030 [**33**]

T. Michael Redmond, Laboratory of Retinal Cell and Molecular Biology, National Eye Institute, National Institutes of Health, Bethesda, MD 20892 [**139**]

Jin-Yuh Shew, Department of Pathology, University of California, San Diego School of Medicine, San Diego, CA 92307 [**221**]

T. Shinohara, Molecular Biology Section, Laboratory of Retinal Cell and Molecular Biology, National Eye Institute, National Institutes of Health, Bethesda, MD 20892 [**163**]

Takashi Shiono, Department of Ophthalmology, Tohoku University School of Medicine, Sendai, Miyagi, Japan [**191**]

Jing-Sheng Si, Laboratory of Retinal Cell and Molecular Biology, National Eye Institute, National Institutes of Health, Bethesda, MD 20892 [**139**]

V.K. Singh, Molecular Biology Section, Laboratory of Retinal Cell and Molecular Biology, National Eye Institute, National Institutes of Health, Bethesda, MD 20892 [**163**]

S. Suzuki, Molecular Biology Section, Laboratory of Retinal Cell and Molecular Biology, National Eye Institute, National Institutes of Health, Bethesda, MD 20892 [**163**]

Judith Toffenetti, Laboratory of Retinal Cell and Molecular Biology, National Eye Institute, National Institutes of Health, Bethesda, MD 20892 [**139**]

Gabriel H. Travis, Department of Psychiatry, University of Texas Southwestern Medical Center, Dallas, TX 75235 [**87**]

M. Tsuda, Molecular Biology Section, Laboratory of Retinal Cell and Molecular Biology, National Eye Institute, National Institutes of Health, Bethesda, MD 20892 [**163**]

Richard G. Weleber, Department of Ophthalmology, Oregon Health Sciences University, Portland, OR 97201 [**191**]

K. Yamaki, Molecular Biology Section, Laboratory of Retinal Cell and Molecular Biology, National Eye Institute, National Institutes of Health, Bethesda, MD 20892 [**163**]

Carmelann Zintz, National Eye Institute, National Institutes of Health, Bethesda, MD 20892; present address: Anatomy Department, Uniformed Services University of the Health Sciences, Bethesda, MD 20814 [**191**]

Preface

This volume, *The Molecular Biology of the Retina: Basic and Clinically Relevant Studies,* is based in part on the proceedings of a symposium that took place in San Francisco at the Eighth Annual Meeting of the International Society of Eye Research in September 1988. Since the molecular biology of the retina is an area of research that has just started to develop and there is not as yet an extensive literature on this subject, we thought that it could be useful for the vision community to gather in the form of a book some of the very interesting data that was discussed at the symposium. In addition to the chapters provided by the speakers, we solicited the contributions of other researchers who are using the tools of molecular biology and genetics for the investigation of basic processes of the normal retina, of animal models of retinal degeneration, and of inherited diseases of the human eye.

The topics covered in this book span a wide range. They include the genetic analysis of mutations affecting the function of the photoreceptor cell; the molecular characterization of the enzymes that hydrolyze cyclic GMP; studies on the genes responsible for two different types of retinal degeneration in mice; the molecular biology of interphotoreceptor retinoid-binding protein, of S-antigen, and the involvement of these proteins in inflammatory diseases of the eye; investigations on the molecular defect in gyrate atrophy; and the recent work on different mutations that cause inactivation of the retinoblastoma gene. All these are areas of great interest in which much progress is being made.

We wish to thank all of the contributors for their time and effort.

Debora B. Farber
Gerald J. Chader

Molecular Biology of the Retina: Basic and Clinically Relevant Studies, pages 1–32

MOLECULAR GENETIC STUDIES OF PHOTORECEPTOR FUNCTION USING DROSOPHILA MUTANTS

William L. Pak

Department of Biological Sciences, Purdue University, West Lafayette, Indiana 47907

INTRODUCTION

The intent of this paper is to review the current status of molecular genetic studies of photoreceptor function that have made use of *Drosophila* mutants. This line of investigation began over twenty years ago when I decided to use *Drosophila* mutants for the express purpose of analyzing the phototransduction process (Pak et al., 1969; Pak, 1975; 1979). The ultimate goal was to understand the photoreceptor mechanism in molecular terms. Although it seemed presumptuous to suggest, at the time, that the use of mutants would eventually lead to molecular analysis of photoreceptor function (see Pak, 1979), this objective is now being realized.

The "molecular genetic approach" that has evolved over the years may be outlined as follows:

1. To isolate, from chemical mutagenesis, mutations that alter light-evoked responses of the eye, ERG (electroretinogram);
2. To identify a subset of mutations that affect phototransduction events;
3. To localize cytogenetically the genes in which the mutations have occurred; and
4. To clone these genes and to analyze them in detail.

In this approach, characterization of mutants (steps 1 & 2) would help determine the functions of the protein products of the genes identified by the mutations, and the cloning and analysis of the genes (steps 3 & 4) would provide molecular information on the proteins. The spectacular recent advances in molecular biology have made cloning and molecular analysis of genes ever easier. However, there have not been comparable advances in the isolation and characterization of mutants, making these the rate-limiting steps in the above approach.

My initial intention was to focus on phototransduction and closely related photoreceptor functions using this approach (see Pak, 1975). Insofar as the genetic approach had not yet been tried in the study of neuronal functions, I felt that it might be more effective when focused on a relatively well-defined problem (Pak, 1975). In the mean time, Dr. S. Benzer and coworkers took a broader perspective and successfully applied this approach to a number of other problems relating to the nervous system (Benzer, 1971; 1973). It soon became clear, however, that many of the mutants that were being isolated on the basis of their defects in the ERG are potentially useful in the dissection of processes other than phototransduction because a defective phototransduction process is not the only cause of ERG defects. The ERG defect-causing mutations that have been isolated fall into nearly 70 complementation groups (Pak, unpublished results). As to be developed further in the last section of this paper, mutations in only a relatively small fraction of these genes appear to affect the transduction process directly. The vast majority of ERG-defective mutants are defective in various other processes carried out by the photoreceptors or the second-order neurons onto which the photoreceptors synapse.

At first sight, one might have thought that the study of phototransduction in Drosophila could turn out to be an esoteric exercise of restricted importance. Evidence accumulating in recent years suggests that it is of much more universal importance than one could have possibly imagined twenty years ago. In the first place, all cells receive signals from outside the cell that are transduced across the cell membrane into intracellular signals to effect various cellular responses (e.g., review by Berridge, 1985). Photoreceptors do not employ a completely new mechanism for phototransduction but a variation of transmembrane signaling found in many other systems (reviewed by Stryer, 1983). Moreover, a growing body of evidence suggests that phototransduction in invertebrate photoreceptors, while similar, is not identical to that in vertebrate photoreceptors. Whereas vertebrate photoreceptors employ a particular variation of transmembrane signaling apparently unique to them, invertebrate phototransduction appears to be based on one of the most commonly employed transmembrane signaling mechanisms (reviews by Berridge, 1985; 1986). Had Drosophila phototransduction been found to be based on an identical mechanism as vertebrate phototransduction, the present approach to phototransduction would have been of considerably diminished significance, since the molecular outline of transduction in vertebrate photoreceptors has been largely elucidated. As it turned out, Drosophila phototransduction is based on a much more universally employed signal transduction mechanism for which molecular details are not yet available, providing us with the opportunity of contributing to one of the central questions in cell biology utilizing the advantages in genetics and molecular biology that Drosophila offers.

The molecular mechanism of phototransduction in vertebrate rods has been largely elucidated in recent years (e.g., reviews by Stryer, 1986; Applebury and Hargrave, 1986). Much less is known about the phototransduction process in invertebrates, although an outline of the process is emerging (Fein, 1986; Payne, 1986; Selinger and Minke, 1988; Bloomquist et al., 1988). As to be developed further later, rhodopsins in the two systems are very similar (O'Tousa et al., 1985; Zuker et al., 1985). In the case of vertebrate photoreceptors, photoexcitation of rhodopsin is coupled to the excitation of a cyclic GMP-dependent phosphodiesterase (cGMP-PDE) through transducin, a class of G protein. The PDE, when activated, hydrolyzes cytoplasmic cGMP, and the decrease in the cytoplasmic cGMP level has been shown to be directly related to the change in conductance of the photoreceptor plasma membrane (Fesenko et al., 1985; Yau and Nakatani, 1985).

Several lines of evidence have led to the hypothesis that, in invertebrate photoreceptors, photoactivated rhodopsin activates a phosphoinositide-specific phospholipase C (PI-PLC) through a G protein (Fein, 1986; Payne, 1986; Devary et al., 1987; Baer and Saibil, 1988; Selinger and Minke, 1988). Activation of PI-PLC hydrolyzes phosphatidylinositol 4,5-bisphosphate (PIP_2) to generate diacylglycerol (DG) and inositol 1,4,5-trisphosphate (IP_3), as in other phosphoinositide-mediated transmembrane signaling systems. Of the two potential second messengers generated, DG and IP_3, IP_3 has been suggested to be the messenger in the ensuing cascade of reactions leading to the generation of the photoreceptor response. The hypothesis is supported by electrophysiological, pharmacological, and biochemical findings. For example, pressure injection of IP_3 into the ventral photoreceptors of Limulus has been shown to mimic the effects of light (Fein et al., 1984; Brown et al., 1984), and light-dependent activation of PI-PLC has been demonstrated in fly and cephalopod eye preparations (Szutz et al., 1986; Devary et al., 1987; Baer and Saibil, 1988). On the other hand, injection of cGMP has also been reported to mimic the effects of light (Johnson et al., 1986), and the possibility of more than one pathway mediating phototransduction events in invertebrates has not yet been excluded.

CHARACTERIZATION OF MUTANTS DEFECTIVE IN PHOTORECEPTOR FUNCTION

The following five genes are considered in this paper: ninaE, norpA, trp, ninaA, and ninaC. All were identified from mutations that affect the ERG. These particular five were chosen because, for one thing, mutations in these genes have been extensively analyzed, making it possible to infer possible functions of the protein products. Moreover, all five genes have now been cloned and molecularly characterized and are known to be expressed specifically in photoreceptors. The proteins encoded by the first two genes, ninaE and norpA, apparently participate directly in the Drosophila phototransduction cascade. The trp mutations also affect phototransduction. However, the role of the trp protein in phototransduction is still unclear. The protein products of the ninaA and ninaC genes are not directly involved in phototransduction. However, they seem to have important roles in the proper organization of either a molecular component of phototransduction or the structural environment needed for phototransduction.

The Gene Encoding R1-6 Opsin, ninaE

The ninaE gene was first identified from mutations that drastically reduce the amount of rhodopsin. To date, ten such genes have been identified. These genes were named ninaA, B, ..., J (neither inactivation nor afterpotential) on the basis of the characteristic electrophysiological phenotype which a reduction in the rhodopsin content confers on the mutants (Pak, 1979; Stephenson et al., 1983) (Fig. 1).

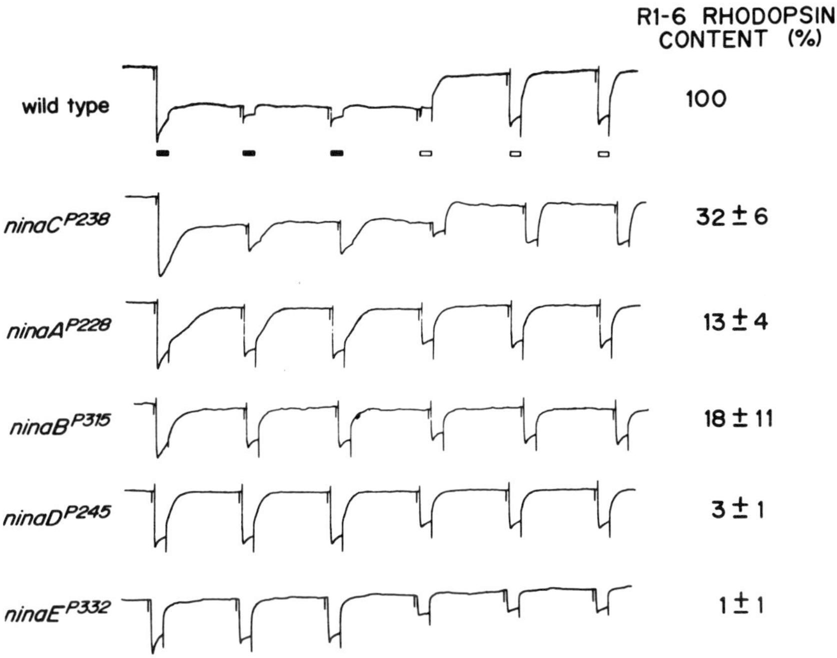

Figure 1. Electroretinograms (ERGs) and relative rhodopsin content obtained from wild-type and one representative mutant from each of five nina complementation groups. The ERGs were obtained by presenting sequentially three blue stimuli (solid bars) followed by three orange stimuli (open bars) at 30-sec intervals. The stimulus duration was 4 sec, and a 5-mV calibration pulse just preceded each response. Shown to the right of each ERG trace is the approximate rhodopsin content in R1-6 photoreceptors for that class of flies, determined by in vivo microspectrophotometry and given as a fraction of the wild-type level. Note the prominent prolonged depolarizing afterpotential (PDA) in wild-type flies; it appears as a response that persists after the cessation of each blue stimulus. The amount of PDA generated by a blue stimulus is correlated with the amount of rhodopsin in R1-6 photoreceptors and decreases from top to bottom in the flies illustrated. All flies used were marked with the mutation white (w) to eliminate the screening pigments in the eye, which otherwise would absorb much of the blue light needed to generate the PDA. Modified from Fig. 1 and Table 2 of Stephenson et al., 1983; reproduced with permission from Cambridge University Press.

The *Drosophila* compound eye contains three classes of photoreceptors, R1-6, R7, and R8. The majority class of photoreceptors, R1-6, contains the same visual pigment, while R7 and R8 contain several spectrally different visual pigments (reviewed by Hardie, 1984). Mutations in two of the *nina* genes examined, *ninaA* and *ninaE*, were found to be R1-6 cell-specific. That is, mutations in these genes reduce only the R1-6 rhodopsin content without any effect on the amount of rhodopsin in either R7 or R8 cells (Stephenson et al., 1983). The R1-6 cell-specificity of *ninaA* and *ninaE* mutations suggested that if the gene encoding R1-6 opsin is represented among the *nina* genes identified, it is very likely either *ninaA* or *ninaE*. To distinguish between these two possibilities, the dependence of the amount of R1-6 rhodopsin on the dosage of normal *ninaA* or *ninaE* allele was examined, since gene dosage dependence is one of the expected characteristics of a structural gene (O'Brien and MacIntyre, 1978). Results showed that the amount of R1-6 rhodopsin covaries with dosage of $ninaE^+$ (Scavarda et al., 1983) but not with that of $ninaA^+$ (Pak et al., 1980), strongly suggesting that the *ninaE* gene encodes R1-6 opsin. The gene was cytogenetically localized to 92B6-7 of the third chromosome using gamma ray induced deficiencies (O'Tousa et al., 1989).

Unequivocal identification of the *ninaE* gene as a rhodopsin gene was obtained from molecular cloning and analysis of the gene (O'Tousa et al., 1985; Zuker et al., 1985). The gene was isolated by taking advantage of its sequence similarity to the bovine rod opsin gene, which had by then been cloned and sequenced (Nathans and Hogness, 1983). Although nucleotide sequence similarity between the *ninaE* and bovine rod opsin genes was relatively low, it was sufficient, nevertheless, to allow cloning of the *ninaE* gene by interspecies hybridization at low stringency. That the *ninaE* gene had been isolated in the cloned DNA fragments was demonstrated by showing 1) that the cloned DNA localized to the *ninaE* (92B6-7) region in *in situ* hybridization to larval salivary chromosomes, 2) that a head-specific 1.7 kb RNA was transcribed from the cloned DNA (O'Tousa et al., 1985; Zuker et al., 1985), and 3) that some of the *ninaE* mutants carried molecular defects in the transcribed region that could be identified in Southern and Northern blot analyses (O'Tousa et al., 1985). The definitive proof that an opsin gene had been cloned, however, was obtained from sequence analysis of the gene.

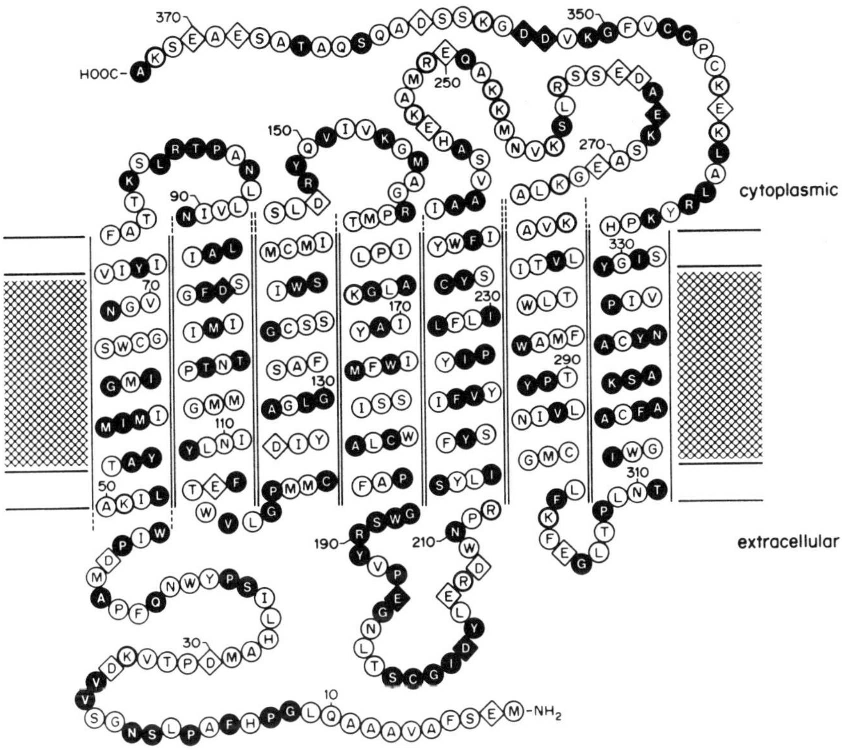

Figure 2. A structural model of Drosophila R1-6 opsin. One letter symbols are used to represent the amino acid sequence. Residues with negative charge are depicted as diamonds, and residues with positive charge are indicated by bold circles. Residues that are identical with those of several mammalian opsin sequences then known are shown in black. For this comparison, Drosophila R1-6 opsin was aligned for maximum homology with complete sequences of human, bovine, and ovine rod opsins and partial sequences of porcine and equine opsins (references found in O'Tousa et al., 1985), introducing insertions and deletions where needed. The seven putative transmembrane segments were identified from hydrophobicity analysis. The lysine residue corresponding to the retinal attachment site is found at position 319 in the seventh transmembrane segment, potential phosphorylation sites represented by five serines and a threonine are found near the C-terminus, and a potential glycosylation site is found at position 20 near the N-terminus. Reproduced from O'Tousa et al., 1985 with permission from Cell.

Sequence analysis of the ninaE gene revealed that the coding sequence predicts a 373-amino acid protein that displays all the known hallmarks of opsin (O'Tousa et al., 1985; Zuker et al., 1985) (Fig. 2). Hydropathy analysis of the amino acid sequence showed that the protein contains seven hydrophobic, presumptive transmembrane domains linked by hydrophilic loops, as was found to be the case for the mammalian opsin sequences that had been elucidated by that time (Ovchinnikov et al., 1982; Hargrave et al., 1983; Nathans and Hogness, 1983, 1984). Moreover, as in the other opsin sequences, the deduced ninaE protein sequence was found to contain the presumed site of chromophore attachment, a lysine residue near the middle of the seventh transmembrane domain; several serine and threonine residues near the C-terminus, thought to be the site of rhodopsin phosphorylation (Hargrave, 1982); and a potential glycosylation site near the N-terminus (Fig. 2). Finally, the *Drosophila* protein showed a great deal of sequence identity with known mammalian opsins. The similarity was particularly striking in several restricted domains, indicating that these domains are likely important for protein function (Fig. 2). These results left little doubt that the ninaE gene encodes a visual pigment protein.

The ninaE sequence was the first invertebrate opsin sequence to be characterized, providing for the first time evidence for the essential identity of molecular architecture between vertebrate and invertebrate opsins. Moreover, by identifying restricted regions of high sequence similarity between mammalian and *Drosophila* opsins, the above work pointed to potential regions of functional importance for opsin proteins. Since then, three other *Drosophila* opsin genes, expressed in other classes of photoreceptors, have been isolated by homology to the ninaE gene (Cowman et al., 1986; Fryxell and Meyerowitz, 1987; Montell et al., 1987; Zuker et al., 1987), and many additional vertebrate and invertebrate opsin sequences have been molecularly characterized. The common molecular features described above have been found in all opsin sequences tested (e.g., review by Pak and O'Tousa, 1988). In fact, many other G protein-coupled membrane receptors, such as adrenergic receptors (Dixon et al., 1986) and muscarinic acetylcholine receptors (Kubo et al., 1986), have also been found to have structural homology with opsin, making opsin a prototype of one of the most widely used receptor proteins in biological systems (e.g., review by Pak and O'Tousa, 1988).

Photoreceptor-Specific Phospholipase C Gene, norpA

Isolation and preliminary characterization of norpA (no receptor potential A) mutants were first reported independently by Pak et al. (1970) and Hotta and Benzer (1970) nearly 20 years ago. These authors showed that norpA mutations, which are sex-linked, drastically affect the light-evoked responses of the eye. Strong mutant alleles of norpA essentially abolish the photoreceptor potential and render the fly blind (reviewed by Pak, 1979). Isolation and identification of a reversibly temperature-sensitive norpA mutant (Deland and Pak, 1973) suggested that norpA is the structural gene for a protein responsible for the no-receptor-potential phenotype. Although it was soon realized that the norpA gene does not encode rhodopsin, nor is it likely to encode components of the ion channels (Pak et al., 1976), the identity of the norpA protein product remained elusive for many years. The first indication that the norpA gene product may somehow be related to phospholipase C was obtained by Yoshioka et al. (1985), who reported that phospholipase C activity, which is normally highly concentrated in the Drosophila eye, is greatly reduced in norpA mutants. Their findings were subsequently confirmed and extended by Selinger and Minke (1988). In particular, the latter investigators showed that, in the temperature-sensitive mutant, PLC activity is normal at permissive temperatures but blocked at restrictive temperatures, indicating that norpA may encode a phospholipase C. In parallel with these findings, a growing body of evidence suggested, though did not prove, that the phototransduction cascade in invertebrate photoreceptors may involve the activation of PLC, as outlined in "Introduction" (references to be found in that section). Moreover, at the time, no phosphoinositide-specific PLC had yet been molecularly characterized in any system. It thus seemed that molecular isolation of the norpA gene might be important for obtaining some critical answers to the problem of invertebrate phototransduction and for better understanding of PLC-mediated transmembrane signaling in general.

Preliminary to cloning, the norpA gene was cytogenetically localized to the interband region between 4B6 and 4C1 near the tip of the X chromosome by deficiency mapping (Banga et al., 1986). The gene was cloned by a combination of chromosomal walking and transposon-mediated mutagenesis (Bloomquist et al., 1988). Chromosomal walking, carried out by H. Steller, C. Montell, and G. Rubin at University of California, Berkeley, yielded two cosmid DNA clones containing approximately 50-kb Drosophila DNA, which mapped to the norpA region defined by deficiency mapping. However, this result alone was not sufficient to be certain that the cloned DNA indeed contained the norpA gene. Fortunately, hobo transposon-mediated mutagenesis (Blackman et al., 1987) yielded a norpA allele with a hobo-element insertion within a piece of DNA corresponding to a small region of the 50-kb DNA cloned into the cosmids. Moreover, remobilization of hobo elements in the germ line of a hobo-induced mutant by suitable dysgenic crosses (Bingham et al., 1982; Blackman et al., 1987) yielded a revertant line among the offspring with normal electrophysiological phenotype and no hobo insertion in the norpA region of the chromosome (G. Koliantz, B. Bloomquist, and R. Shortridge, unpublished results), strongly indicating that insertion of a hobo element into the norpA gene was responsible for the norpA phenotype in the hobo-induced mutant. These results demonstrated that the norpA gene is indeed contained within the cloned boundaries of the 50-kb DNA. Moreover, the location of the gene within the 50-kb region could now be identified from the site of hobo insertion.

Preliminary determination of the extent of the norpA gene took advantage of the fact that the gene is likely to be expressed predominantly, if not exclusively, in the eye (hence, in the head), since mutations in this gene affect only the eye. Hybridization of DNA fragments from the cloned 50-kb region to head or body RNA blots identified a region that encodes a 7.5 kb RNA expressed only in the head. The site corresponding to hobo insertion site in the hobo-induced mutant was found to be within this region, consistent with the supposition that this region corresponds to the norpA gene. DNA fragments from this head-specific region were used as probes to isolate cDNAs, and several representative cDNAs were subjected to sequence analysis, revealing the presence of a single open reading frame in the largest cDNAs isolated. Tissue distributions of gene products were determined by in situ hybridization of radiolabeled RNA probes made from one of the cDNAs to tissue sections and immunolocalization of the protein product using antisera against a lacZ-norpA fusion protein. Results of in situ hybridization showed heaviest labeling in the retina, and the antisera recognized the rhabdomeres of photoreceptors specifically (S. Schneuwly and M. Perdew, unpublished results).

Conceptual translation of the open reading frame generated a polypeptide of 1095 amino acids. Although no PI-PLC sequence was available at the time the norpA sequence was first being generated, a number of mammalian PI-PLC sequences became available from cDNA cloning shortly thereafter. Of particular interest for the present discussion are the deduced amino acid sequences for three immunologically distinct PI-PLCs that were first isolated from the bovine brain: PLC-I (PLCβ: Katan et al., 1988; Suh et al., 1988a), PLC-II (PLCγ: Stahl et al., 1988; Suh et al., 1988b), and PLC-III (PLCδ: Suh et al., 1988a). Suh et al. (1988a) noted that the overall sequence similarity among these proteins is not all that high, but strikingly high sequence similarities are found in restricted domains, particularly two regions of about 120 to 180 amino acids each (44-61% identity), which they now call Region X and Region Y (Rhee et al., 1989). Comparison of the norpA sequence with these mammalian sequences showed that it displays strong sequence similarity with all three mammalian proteins. The similarity was particularly high between the norpA protein and PLC-I (36% identity overall; 60 and 58% identity in Regions X and Y, respectively) (Fig. 3). In fact, mammalian PLC-I was found to show greater sequence similarity with the deduced norpA protein than with either PLC-II or PLC-III from the same tissue of the same species. Although we have not yet expressed the norpA protein to directly demonstrate PLC activity, the above results strongly indicated that the norpA gene encodes a PI-PLC.

Since in the absence of the norpA gene function little or no receptor potential is generated, it was concluded that the photoreceptor-specific PI-PLC encoded by the norpA gene is an integral component of the Drosophila phototransduction pathway. This conclusion, of course, does not exclude the possibility of multiple, synergistic pathways contributing to phototransduction following PLC activation. Even the possibility of another pathway operating in parallel with the PLC pathway cannot yet be eliminated, although such a pathway, if it exists at all, is likely to make only a minor contribution.

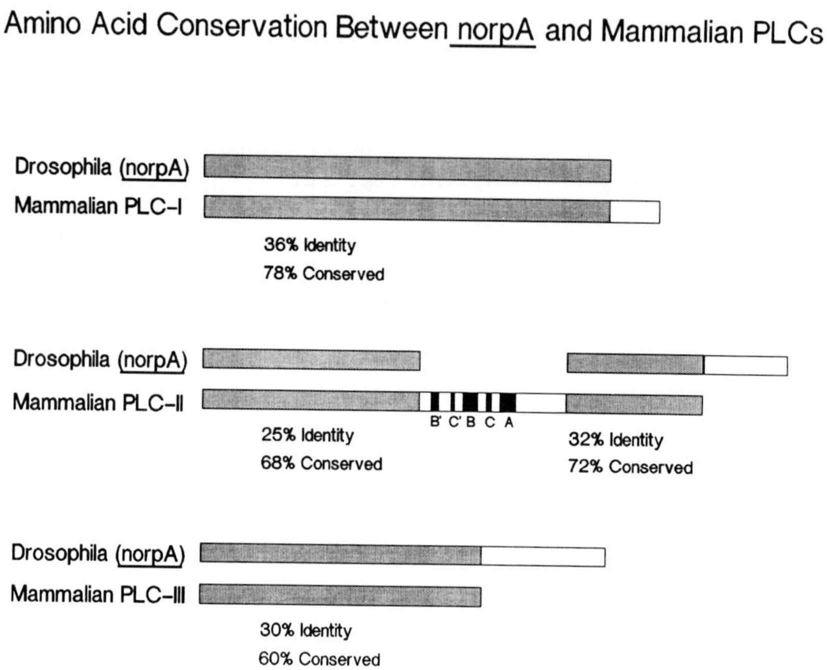

Figure 3. Graphic representation of sequence similarity between the norpA protein and phosphoinositide-specific phospholipase C's (PI-PLCs) found in mammalian brain. For each of the three cases, comparisons were made over the shaded regions of proteins. "Percent identity" represents the fraction of identical amino acids found, and "percent conserved" includes conservative substitutions (Dayhoff, 1978). The *Drosophila* protein bears greatest sequence similarity to PLC-I (PLC-β). In fact, PLC-I is more similar to the norpA protein than to PLC-II (PLC-γ) or PLC-III (PLC-δ) of the same species. PLC-II contains sequences that are not found in other PLCs in the central region of the protein, and it was necessary to introduce a large gap into the norpA protein for comparison with this protein. The central portion of the figure comparing norpA protein with PLC-II is reproduced from Bloomquist et al., 1988 with permission from Cell.

The trp Gene

Unlike the other genes discussed in this paper, the trp gene was first identified from a spontaneously arising mutant that behaved abnormally when the ambient light level was raised (Cosens and Manning, 1969; Cosens, 1971) and not from chemical mutagen-generated ERG-defective mutants. A chemical mutagenesis program, however, did generate subsequently several mutant alleles at this locus (Pak, 1979). Most of the work on the trp gene was carried out in other laboratories. Nevertheless, trp is included in this paper because it is one of the most extensively analyzed loci and because mutations at this locus do affect the photoreceptor response.

Cosens and Manning (1969) and Cosens (1971) showed that the electroretinogram (ERG) taken from this mutant, while normal at low stimulus intensities, decays to baseline at high stimulus levels while the stimulus is still on, providing a likely explanation for the "blindness" of the mutant at high light levels. Minke et al. (1975) demonstrated by means of intracellular recordings that the photoreceptor potential indeed decays during a bright and/or prolonged light stimulus. In addition to the rapid decay during stimulus, the recovery of the receptor potential following strong adapting light is also abnormally slow in the mutant (Cosens, 1971; Minke, 1982), suggesting possible involvement of the adaptation process. Minke (1982) showed, however, that the above mutant phenotypes are not simply manifestations of unusually strong adaptation to light. For one thing, a previous exposure to adapting light greatly increases the latency and rise time of test responses in the mutant, while the normal effects of previous adaptation on test responses are to decrease these parameters. Moreover, noise analysis revealed that the individual miniature potentials, or bumps, which are the elementary units of the receptor potential, do not simply display exaggerated light adaptation when the mutants are exposed to bright light (Minke et al., 1975).

It has been recognized since the sixties that the receptor potential is generated by a summation of a large number of bumps, each of which arises in response to a single photon (Fuortes and Yeandle, 1964; Rushton, 1961; Dodge et al., 1968). With increasing stimulus intensities, the average bump rate increases and the bump amplitude decreases when measured at the steady-state phase of the receptor potential (Dodge et al., 1968; Wu and Pak, 1978 for Drosophila). This decrease in the bump amplitude at steady state with high light stimulus is thought to be responsible for the characteristic decay of the receptor potential from its initial peak to the steady-state level during light stimulation (Dodge et al., 1968). If the decay of the _trp_ receptor potential to baseline during light is an exaggerated form of adaptation normally seen in wild type, one would expect the bump amplitude to decrease even more strongly with increasing light intensities at steady state in _trp_ than in wild type, while the bump rate would increase as in wild type. Contrary to this expectation, noise analysis (Minke et al., 1975) of _trp_ responses revealed that the average bump amplitude decreases very little with light intensities. Moreover, the average bump rate at steady state remains low at high light intensities. Thus, it was concluded that the decay of the receptor potential in _trp_ mutants arises not because the bumps adapt to abnormally low values but because the efficiency of bump generation becomes greatly reduced (Minke et al., 1975; Minke, 1982).

Wong et al. (1989) localized the _trp_ gene to 99C5-6 bands at the tip of the right arm of the third chromosome using a set of chromosome duplications (Frisardi and MacIntyre, 1984). Meanwhile, Levy et al. (1982) had isolated a set of 20 genomic clones containing sequences that are expressed more abundantly in the head than body and showed that one of these clones, λ559, mapped to 99C. The _trp_ gene was isolated using this clone (Montell et al., 1985; Montell and Rubin, 1989; Wong et al., 1989). Because RNA blot studies had indicated that the λ559 clone did not contain the entire _trp_ coding region, Montell et al. (1985) isolated an overlapping clone from a _Drosophila_ genomic library using λ559 as a probe. This λ559-derived clone was then used to rescue a _trp_ mutant by P-element-mediated germ line transformation (Rubin and Spradling, 1982), providing definitive evidence that the _trp_ gene is contained within this clone. Wong et al. (1989), meanwhile, isolated the _trp_ gene by chromosomal walking that started with the λ559 clone.

The results of the two groups are very similar, though not identical (Montell and Rubin, 1989; Wong et al., 1989). The gene structure (14 exons and 13 introns), the size of the transcript [4.1 kb (Montell and Rubin, 1989); 4.5 kb (Wong et al., 1989)], the deduced protein structure, and the size of the protein product [1275 amino acids (Montell and Rubin, 1989); 1274 amino acids (Wong et al., 1989)] reported by the two groups are all very similar. The deduced trp protein shows no sequence similarity to any known protein. The protein may be divided into several domains according to their hydrophobicity. The most hydrophobic domain is found in the central region while the carboxyl regions are generally hydrophilic, particularly near the C-terminus. Found near the C-terminus is an eight amino acid sequence, D-K-D-K-K-P-G/A-D, repeated nine times (Montell and Rubin, 1989). Montell and Rubin have found, on the basis of hydrophobicity analysis, that there are four, and possibly as many as eight, segments in the central hydrophobic domain that satisfy the Kyte and Doolittle (1982) criteria for transmembrane segments and concluded that the trp protein is an integral membrane protein. Wong et al. (1989), on the other hand, have carried out secondary structure analysis in addition to hydrophobicity analysis and looked for stretches that are both helical and hydrophobic. They came to the conclusion that the protein does not "have hydrophobic helices of the length required" to span the membrane, though it could "be a membrane-associated protein." Immunocytochemical localization of the protein using antibodies raised against a synthetic peptide (Wong et al., 1989) or a β-galactosidase-trp fusion protein (Montell and Rubin, 1989) has demonstrated that the protein is localized in the photoreceptors, indeed, very probably in the rhabdomeres. However, the identity of the protein remains unknown.

The ninaA Gene

Mutations in the remaining two genes to be discussed, ninaA and ninaC, do not affect the phototransduction process directly (ninaA: Larrivee et al., 1981; ninaC: Matsumoto et al., 1987). Rather, their effects are on R1-6 rhodopsin and the cytoskeleton of microvilli comprising the rhabdomeres, respectively.

Several lines of evidence, including freeze-fracture electron microscopy, had established that ninaA mutations specifically reduce the amount of R1-6 rhodopsin (Larrivee et al., 1981; Schinz et al., 1982; Stephenson et al., 1983). Thus, of the ten nina genes that have been identified to date (see section on the ninaE gene), so far mutations in two of them, ninaA and ninaE, have been found to reduce the amount of R1-6 rhodopsin without affecting either the R7 or R8 rhodopsin content. Mutations in the ninaE gene were expected to be specific for R1-6 rhodopsin because the gene encodes R1-6 opsin (O'Tousa et al., 1985; Zuker et al., 1985). An intriguing question was why ninaA mutations are also specific for R1-6 rhodopsin. We had thought initially that ninaA might be a regulatory gene for ninaE, the R1-6 opsin gene. It has been shown, however, that ninaA mutations have no effect either on the ninaE transcription level or ninaE mRNA stability (Zuker et al., 1988), making regulation of ninaE by ninaA unlikely, at least at the transcription level. Molecular cloning of the ninaA gene was undertaken in the hope of identifying its protein product and uncovering its relationship to R1-6 rhodopsin.

Our group used deficiency mapping to localize the ninaA gene to the 21D1-2 to 21E1-2 region of the second chromosome (Larrivee et al., 1981; Schneuwly et al., 1989) and isolated it in a chromosomal walk that covered the entire chromosomal region defined by the mapping (Schneuwly et al., 1989). Independently, C. Zuker's group at University of California at San Diego also cloned this gene by making use of a pool of "eye-specific" genomic clones that they had generated (Shieh et al., 1989). Though they began their work after we did, they cloned the ninaA gene and identified the protein product somewhat before we did and kindly made their results known to us. The following is an account of our work (Schneuwly et al., 1989); similar work was also carried out independently by the above group (Shieh et al., 1989).

Among the overlapping genomic fragments isolated in the chromosomal walk, that containing the ninaA gene was identified by looking for those containing sequences that are transcribed specifically in the eye, since ninaA mutations affect only the eye. Within the chromosomal region covered by the walk, only one restriction fragment was found to satisfy this criterion, a 1.6 kb BglII fragment. Sequence analysis of the fragment and cognate cDNAs revealed a single open reading frame with coding capacity of 237 amino acids. To prove that the ninaA gene had been cloned, several known ninaA mutants were examined to see if the candidate gene contains nucleotide changes that would alter the translated product in the mutants. Sequence analysis of the same 1.6 kb BglII fragments isolated from three different EMS (ethyl methanesulfonate)-

induced ninaA mutants showed that each of the three alleles contains a single base change that would alter the translated product, in strong support of the conclusion that the ninaA gene had been cloned.

Comparison of the ninaA sequence with those in DNA and protein data bases revealed strong sequence similarity between the ninaA-encoded protein and cyclophilin (CyP), a cyclosporine A (CsA) binding protein (Shieh et al., 1989; Schneuwly et al., 1989). CsA is a cyclic undecapeptide of fungal origin extensively used in human organ transplantation because of its potent immunosuppressive properties (Borel, 1986; Beveridge, 1986). Unlike most CyPs analyzed to date, the ninaA protein appears to be a membrane bound-protein. The ninaA encoded protein is larger than other CyPs (e.g., human and *Neurospora* CyPs: Schneuwly et al., 1989) and contains in both the N-and C-terminal regions hydrophobic domains which could serve as a signal sequence and a membrane anchor, respectively (Schneuwly et al., 1989). In spite of the topological differences between the ninaA encoded protein and the other CyPs, the ninaA protein has been shown to be a CsA-binding protein (Shieh et al., 1989; Schneuwly et al., 1989).

Cellular functions of CyP have not been elucidated, although it has been suggested to play a role in a signaling pathway associated with T cell activation (Harding and Handschumacher, 1988). We therefore examined the ninaA mutants electrophysiologically to see if there is any evidence that the phototransduction pathway may be affected (Schneuwly et al., 1989). One of the mutants examined is almost certain to be a null mutant because the ninaA mRNA level is undetectable in RNA blot analysis and the mutant carries a nonsense mutation which would allow only a small truncated protein product to be generated. Even this mutation had no obvious effect on the physiological responses of R1-6 photoreceptors except those that could be readily accounted for by the reduced rhodopsin content (Fig. 4), suggesting that the ninaA protein is not directly involved in phototransduction. The only detectable mutant phenotype that correlates with the severity of molecular defects in the three ninaA mutants examined is the amount of depletion of R1-6 rhodopsin (Schneuwly et al., 1989) (Fig. 4). Recently, it has been shown that CyP is identical to peptidyl prolyl *cis*-*trans* isomerase (Fischer et al., 1989; Takahashi et al., 1989), implicated in protein folding

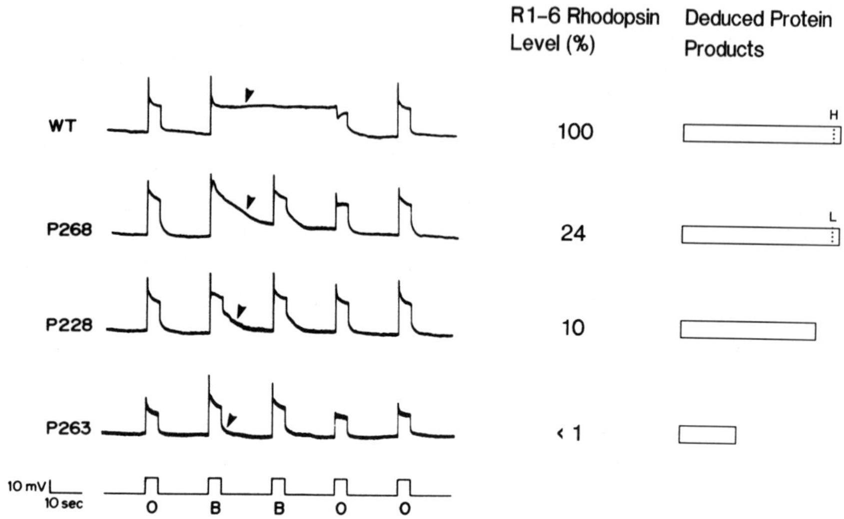

Figure 4. Electrophysiological phenotypes, R1-6 rhodopsin levels, and bar graph representations of deduced protein products of wild type and three allelic mutants of ninaA. Intracellular recordings of the receptor potentials were obtained from R1-6 photoreceptors of wild type (first trace on the left), ninaA^{P268} (second trace), ninaA^{P228} (third trace), ninaA^{P263} (fourth trace). The responses were elicited by a series of orange (O) and blue (B) stimuli of 4-sec durations, as shown in the bottom trace on the right. All flies were marked with the mutation white (w) to eliminate the eye color pigments. The PDA generated by the first blue stimulus is indicated by an arrowhead in each trace. R1-6 rhodopsin levels were determined by *in vivo* microspectrophotometry. The amount of PDA generated co-varies with the amount of R1-6 rhodopsin present, though nonlinearly. The amino acid sequences of protein products were deduced from DNA sequencing. The ninaA^{P268} mutant allele carries a missense mutation that alters the histidine residue at position 227 into leucine; while the ninaA^{P263} and ninaA^{228} alleles carry nonsense mutations that truncate the putative protein at positions 87 and 208, respectively. The only observable mutant phenotype that correlates with the severity of molecular defects in the three ninaA mutants is the amount of depletion of R1-6 rhodopsin, which, in turn, is responsible for the PDA phenotype.

(Lang et al., 1987). Thus, the role of the ninaA protein may be in the proper folding of R1-6 opsin. Since R1-6 opsin is an integral membrane protein, it may require a membrane-bound version of CyP for its folding.

That CyP plays a role at all in rhodopsin function came as a big surprise. The accumulated knowledge on rhodospin from over a century of research on vertebrate rhodopsin has provided no inkling that this might be the case. Yet, this may be only the first of a number of such proteins the present mutant approach to photoreceptor function may uncover in the future. Moreover, the availability of ninaA mutants may prove useful in further investigation of the role of CyP in cell physiology.

The ninaC Gene

In comparison to other classes of nina mutations studied, the effects of ninaC mutations on the rhodopsin content are relatively mild, decreasing it to about 25 to 40% of the wild-type level in young (<7 days post-eclosion) flies (Stephenson et al., 1983; Matsumoto et al., 1987). By contrast, ninaA mutations decrease the R1-6 rhodopsin content to 10% or less, depending on the allele examined (L. L. Randall, unpublished results), and ninaE mutations decrease it to 1% or less (Johnson and Pak, 1986). On the other hand, for a given decrease in rhodopsin content, a ninaC mutation would decrease the size of the rhabdomeres much more severely than other nina mutations studied to date (Matsumoto et al., 1987). Histological reconstruction of rhabdomeres of several ninaC mutants suggested that the decrease in their size most likely accounts for most, if not all, of the observed decrease in rhodopsin content (K. Isono, unpublished data). In support of this view, ultrastructural observations of transverse sections through rhabdomeral microvilli of several ninaC mutants showed that the electron dense core normally present within each microvillus is greatly reduced or missing in the mutants (Matsumoto et al., 1987) (Fig. 5). Previous investigators working with other invertebrates have shown that these densities are associated with the cytoskeleton of the rhabdomeral microvillus (Saibil, 1982; Blest et al., 1982). Thus, our results led to the interpretation that the ninaC gene encodes components of the cytoskeleton of rhabdomeral microvilli and that, in ninaC mutants, defects in these

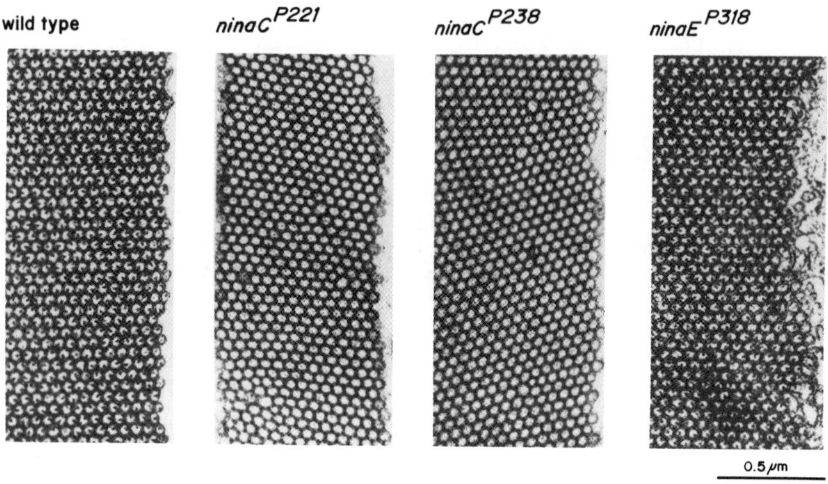

Figure 5. Electron micrographs of transverse sections through rhabdomeral microvilli of R1-6 photoreceptors of wild type, <u>ninaC</u>^{P221}, <u>ninaC</u>^{P238}, and <u>ninaE</u>^{P318}. Note the prominent electron-dense axial cores in the microvilli of wild type and <u>ninaE</u>^{P318}, but not in those of the two <u>ninaC</u> mutants. The <u>ninaE</u>^{P318} mutant carries a defect in the R1-6 rhodopsin gene, <u>ninaE</u>, and has greatly reduced R1-6 rhodopsin content. This mutant is included as a control to indicate that a reduction in the rhodopsin content does not affect the axial cores of the rhabdomeral microvilli. Only young flies, 2 to 5 days post-eclosion, were used. Reproduced from Matsumoto et al., 1987 with permission from Proceedings of National Academy of Sciences.

---

cytoskeletal components cause the rhabdomeres to fall apart, leading to reductions in the amount of rhodopsin. Unlike <u>ninaA</u> or <u>ninaE</u> mutations, the effect of <u>ninaC</u> mutations on the rhodopsin content is, therefore, indirect.

Information on the <u>ninaC</u>-encoded protein products were obtained from SDS polyacrylamide gel electrophoresis (SDS PAGE) of eye tissues (Matsumoto et al., 1987) and molecular isolation and analysis of the gene (Montell and Rubin, 1988). Comparative SDS-PAGE analyses of microdissected eye and head tissues identified five eye-specific protein bands (designated B1 to B5) that ranged in apparent molecular mass from 170 kD for B1 to 120 kD for B5. Although ninaC mutations affect all five bands, their effects were particularly strong on B1 and B4. Many of the ninaC alleles tested either eliminated or altered the mobility of one or both of these bands. Moreover, when segmental aneuploid flies carrying varying copies of $ninaC^{+}$ were constructed and tested, the amount of B1 and B4 were found to vary with $ninaC^{+}$ in a dosage-dependent manner. These results led to the suggestion that the ninaC gene encodes B1 and B4 polypeptides (Matsumoto et al., 1987).

The ninaC gene was isolated (Montell and Rubin, 1988) using the "head enriched" genomic clones which Levy et al. (1982) had previously isolated for another purpose (see the section on the trp gene). Levy et al. (1982) had shown that one of their clones, λ551, mapped to 28A, within the interval 27E-28B on the second chromosome, to which we had mapped the ninaC gene (Matsumoto et al., 1987). Montell and Rubin (1988) used a DNA fragment derived from λ551 clone to rescue a ninaC mutant by P-element-mediated germ line transformation (Rubin and Spradling, 1982; Spradling and Rubin, 1982), providing proof that the gene is contained within the cloned DNA fragment.

RNA blot analysis showed that two species of RNA, about 4 kb and 5 kb, are expressed from the ninaC locus (Montell and Rubin, 1988) beginning in the late pupal stage and continuing in the adult. Similar patterns of RNA expression are also found for all other eye-specific genes tested. Amino acid sequences of the proteins encoded by these mRNAs were deduced from cDNA sequencing (Montell and Rubin, 1988). The sequence analysis showed that the two species of mRNAs encode two closely related proteins of 1135 and 1501 amino acids (132 and 174 kD). The N-terminal 1081 of these amino acids were found to be common to both proteins. Contained within the common N-terminal region of the proteins are two adjacent domains of 266 and 725 amino acids that are homologous to protein kinases and the head portion of the myosin heavy chain, respectively. The C-terminal regions of the two proteins, however, did not show any significant sequence similarity to any previously sequenced proteins or to each other. Polyclonal antiserum prepared against a lacZ-ninaC fusion protein were used to identify ninaC protein products in Western blots of SDS-PAGE separated Drosophila eye proteins and to immunolocalize the ninaC proteins in tissue sections (Montell and Rubin, 1988). The results showed that the antiserum recognizes the two protein bands, B1 and B4, we had previously identified to be ninaC proteins from genetic evidence (Matsumoto et al., 1987). Moreover, these authors showed that the expression of ninaC proteins is restricted to the rhabdomeres of the compound eyes and the ocelli. These results are consistent with our previous suggestion that the ninaC proteins are components of the cytoskeletal structure of the rhabdomeres. However, the fact that these proteins contain a domain homologous to protein kinases suggests that they may also have functions beyond purely structural ones (Montell and Rubin, 1988).

POTENTIAL FUTURE DIRECTIONS

Of the nearly 70 genes that have so far been identified to cause defects in the ERG when a mutation is induced in the gene, several have now been characterized both at the genetic and molecular levels, as described above. Several others are now being isolated either in this laboratory or elsewhere. Information obtained from the small number of genes that has been analyzed to date has already proved to be far more valuable than one could have imagined only a few years ago. The future progress in this line of research, however, will depend on identification, isolation, and characterization of additional genes that encode proteins involved in phototransduction and other functions carried out by the photoreceptor. As outlined in "Introduction," the approach that we have taken so far has been to analyze the mutants in some detail first so as to determine which mutants might be more "interesting" or "important" before proceeding to molecular isolation of the genes that are defective in the mutants. As valuable as this approach has been to date, it will not be efficient enough to cope with the large number of genes that need to be analyzed in a reasonable time frame. Moreover, it is now becoming increasingly clear that any gene which, when defective, causes ERG defects is likely to be important for some photoreceptor function. Thus, it seems worth considering molecular isolation and subsequent analysis of all, or a large fraction, of the genes identified by ERG-defect causing mutations using more efficient cloning strategies, regardless of whether or not a detailed analysis has been done on mutations in the genes. Mutational analysis would then be carried out along with molecular analysis.

The cloning strategy of choice appears to be the use of eye-specific clones, i.e., genomic or cDNA clones that contain sequences expressed preferentially in the eye. If the genes that have been isolated to date are any guide, most genes important in photoreceptor function apparently are expressed specifically, or preferentially, in the photoreceptors (O'Tousa et al., 1985; Zuker et al., 1985; Bloomquist et al., 1988; Montell and Rubin, 1988, 1989; Van Vactor et al., 1988; Shieh et al., 1989; Schneuwly et al., 1989; Wong et al., 1989). Providing that mutations in such eye-specific genes are responsible for most of the ERG-defective mutants that have been isolated, the chromosomal positions of many eye-specific clones should match those of ERG-defective mutants, if a large enough pool of eye-specific clones is generated and tested. Confirmation that the gene responsible for the ERG defect in a mutant is contained within the eye-specific clone identified from its chromosomal position can then be made by, for example, *in vivo* complementation of mutant phenotype by P-element-mediated transformation (Rubin and Spradling, 1982) of the mutant germ line with the cloned DNA.

The basic idea outlined above has already been successfully utilized. Pools of eye-specific clones have been generated by previous workers (Levy et al., 1982; Palazzolo et al., 1987; Shieh et al., 1989), and the use of these clones has been very effective in isolating such genes as *trp* (Montell et al., 1985), *ninaC* (Montell and Rubin, 1988), and *ninaA* (Shieh et al., 1989), as described above. More recently, C. Zuker's group at University of California at San Diego has also cloned an eye-specific protein kinase C gene (Schaeffer et al., 1989) and a *Drosophila* homolog of the S-antigen gene (C. Zuker, personal communication), neither of which was identified previously by mutations. Implementation of the above strategy, however, will require not only that a large pool of eye-specific clones be made available, but also, even more importantly, that more accurate mapping data be generated on the available ERG-defective mutants. Most of the ERG-defective mutants isolated remain largely uncharacterized. A relatively small number of genes corresponding to the mutants that have been well characterized either have already been cloned or are being cloned. In particular, reliable deletion mapping has not yet been carried out on the great majority of these mutants. However, preliminary classification of many of the mutants is possible from the available data. The classes of mutants available, in turn, could suggest possible areas in which the present line of research could make future contributions.

Among the more interesting classes of mutants are two highly overlapping ones consisting of mutants with small or no ERG and those displaying photoreceptor degeneration with age. Each class consists of over a dozen complementation groups, if counted separately. In those mutants that belong to both classes, the mutant ERG phenotype almost always appears long before any photoreceptor degeneration is apparent, suggesting that photoreceptor degeneration is not responsible for the degraded ERG response. Thus, many of the mutants in these classes probably are defective in the phototransduction cascade. Moreover, the degeneration mutants are of intrinsic interest in themselves because of the prevalence of photoreceptor degeneration as a cause of inherited blindness in the human (e.g., Berson, 1977). Another class of note is the nina class consisting of mutants with greatly reduced R1-6 rhodopsin content (see discussion in the section on the ninaE gene). These mutants fall into 10 complementation groups. The ninaA, C, and E genes discussed above all belong in this class. Detailed characterization of the remaining nina genes likely will contribute to further understanding of rhodopsin-related functions. Still another class consists of mutants in a dozen complementation groups, all of which share the phenotype that the time course of decay of the ERG is greatly increased or decreased. Some of these mutants probably carry defects in ion channels that shape the waveform of the receptor potential. By far the largest class (nearly 30 complementation groups) consists of mutants in which the on- and off-transients of the ERG is reduced or missing. Recently, Coombe (1986) has provided evidence that these transients are generated by the large monopolar neurons in the lamina. These are the neurons onto which the photoreceptors synapse. Thus, the defects in many of the mutants in this class could be due to the absence of signal transmission between the photoreceptors and the second-order neurons in the lamina. For example, defects in transmitter synthesis or release, the receptor molecule in the postsynaptic membrane, or transmembrane signaling in the second-order neuron could all contribute to the mutant phenotype. The defects suggested by these mutants enumerated above are then some of the areas in which the use of ERG-defective mutants is likely to make contributions in the future.

ACKNOWLEDGMENTS

The author thanks J. Bentrop, L. L. Randall, S. Schneuwly, and R. D. Shortridge for helpful comments and discussions and B. A. Petty for help with the preparation of the manuscript. The work in this laboratory was supported by grants from the National Eye Institute (R01-EY00033, R01-EY04767, P30-EY02723) and the National Science Foundation.

REFERENCES

Applebury ML, Hargrave PA (1986). Molecular biology of the visual pigments. Vision Res 26:1881-1895.

Baer KM, Saibil HR (1988). Light- and GTP-activated hydrolysis of phosphatidylinositol bisphosphate in squid photoreceptor membranes. J Biol Chem 263:17-20.

Banga SS, Bloomquist BT, Brodberg RK, Pye QN, Larrivee DC, Mason JM, Boyd JB, Pak WL (1986). Cytogenetic characterization of the 4BC region on the X chromosome of Drosophila melanogaster: Localization of the mei-9, norpA and omb genes. Chromosoma 93:341-346.

Benzer S (1971). From gene to behavior. J Am Med Assoc 218:1015-1022.

Benzer S (1973). Genetic dissection of behavior. Sci Am 229:24-37.

Berridge MJ (1985). The molecular basis of communication within the cell. Sci Am 253:142-152.

Berridge MJ (1986). Inositol phosphates as second messengers. In Putney JW Jr (ed): "Phosphoinositides and Receptor Mechanisms," New York: Alan R. Liss, pp 25-45.

Berson EL (1977). Hereditary retinal diseases: classification with the full-field electroretinogram. In Lawwill T (ed): "Doc Ophthalmol Proc Ser, Vol 13: ERG, VER, and Psychophysics" (XIVth ISCERG Symposium, May 1976, Louisville, Ky), The Hague, The Netherlands: Dr. W. Junk, pp 149-171.

Beveridge T (1986). Clinical transplantation - overview. In Borel JF (ed): "Progress in Allergy, Vol 38: Ciclosporin," Basel: Karger, pp 269-292.

Bingham PM, Kidwell MG, Rubin GM (1982). The molecular basis of P-M hybrid dysgenesis: the role of the P element, a P-strain-specific transposon family. Cell 29:995-1004.

Blackman RK, Grimaila R, Koehler MMD, Gelbart WM (1987). Mobilization of hobo elements residing within the decapentaplegic gene complex: suggestion of a new hybrid

dysgenesis system in Drosophila melanogaster. Cell 49: 497-505.
Blest AD, Stowe S, Eddey W (1982). A labile, Ca^{+2}-dependent cytoskeleton in rhabdomeral microvilli of blowflies. Cell Tissue Res 223:553-573.
Bloomquist BT, Shortridge RD, Schneuwly S, Perdew M, Montell C, Steller H, Rubin G, Pak WL (1988). Isolation of a putative phospholipase C gene of Drosophila, norpA, and its role in phototransduction. Cell 54:723-733.
Borel JF (1986). Editorial: Ciclosporin and its future. In Borel JF (ed): "Progress in Allergy, Vol 38: Ciclosporin," Basel: Karger, pp 9-18.
Brown JE, Rubin LJ, Ghalayini AJ, Tarver AP, Irvine RF, Berridge MJ, Anderson RE (1984). myo-inositol polyphosphate may be a messenger for visual excitation in Limulus photoreceptors. Nature 311:160-163.
Coombe PE (1986). The large monopolar cells L1 and L2 are responsible for ERG transients in Drosophila. J Comp Physiol 159:655-665.
Cosens D (1971). Blindness in a Drosophila mutant. J Insect Physiol 17:285-302.
Cosens D, Manning A (1969). Abnormal electroretinogram from a Drosophila mutant. Nature 224:285-28.
Cowman AF, Zuker CS, Rubin GM (1986). An opsin gene expressed in only one photoreceptor cell type of the Drosophila eye. Cell 44:705-710.
Dayhoff M (1978). "Atlas of Protein Sequence and Structure, Volume 5, Supplement 3." Silver Spring, Maryland: National Biomedical Research Foundation.
Deland MC, Pak WL (1973). Reversibly temperature sensitive phototransduction mutant of Drosophila melanogaster. Nature New Biol 244:184-186.
Devary O, Heichal O, Blumenfeld A, Cassel D, Suss E, Barash S, Rubinstein CT, Minke B, Selinger Z (1987). Coupling of photoexcited rhodopsin to inositol phospholipid hydrolysis in fly photoreceptors. Proc Natl Acad Sci USA 84:6939-6943.
Dixon RAF, Kobilka BK, Strader DJ, Benovic JL, Dohlman HG, Frielle T, Bolanowski MA, Bennett CD, Rands E, Diehl RE, Mumford RA, Slater EE, Sigal IS, Caron MG, Lefkowitz RJ, Strader CD (1986). Cloning of the gene and cDNA for mammalian β-adrenergic receptor and homology with rhodopsin. Nature, Lond 321:75-79.
Dodge FA Jr, Knight BW, Toyoda J (1968). Voltage noise in Limulus visual cells. Science 160:88-96.
Fein A (1986). Excitation and adaptation of Limulus

photoreceptors by light and inositol 1,4,5-trisphosphate. Trends in Neurosciences 9:110-114.

Fein A, Payne R, Corson DW, Berridge MJ, Irvine RF (1984). Photoreceptor excitation and adaptation by inositol 1,4,5-trisphosphate. Nature 311:157-160.

Fesenko EE, Kolesnikov SS, Lyubarsky AL (1985). Induction by cyclic GMP of cationic conductance in plasma membrane of retinal rod outer segment. Nature, Lond 313:310-313.

Fischer G, Wittmann-Liebold B, Lang K, Kiefhaber T, Schmid FX (1989). Cyclophilin and peptidyl-prolyl cis-trans isomerase are probably identical proteins. Nature 337:476-478.

Frisardi MC, MacIntyre RJ (1984). Position effect variegation of an acid phosphatase gene in Drosophila melanogaster. Mol Gen Genet 197:403-413.

Fryxell KJ, Meyerowitz EM (1987). An opsin gene that is expressed only in the R7 photoreceptor cell of Drosophila. EMBO J 6:443-451.

Fuortes MGF, Yeandle S (1964). Probability of occurrence of discrete potential waves in the eye of Limulus. J Gen Physiol 47:443-463.

Hardie RC (1984). Functional organization of the fly retina. In Autrum H, Ottoson D, Perl ER, Schmidt RF, Shimazu H, Willis WD (eds): "Progress in Sensory Physiology 5," Berlin: Springer-Verlag, pp 1-79.

Harding MW, Handschumacher RE (1988). Cyclophilin, a primary molecular target for cyclosporine. Transplantation 46:29S-35S.

Hargrave PA (1982). Rhodopsin chemistry, structure and topography, In Osborne N, Chader G (eds): "Progress in Retinal Research," Oxford: Oxford University Press, pp 1-51.

Hargrave PA, McDowell JH, Curtis DR, Wang JK, Juszczak E, Fong S-L, Mohanno Rao JK, Argos P (1983). The structure of bovine rhodopsin. Biophys Struct Mech 9:235-244.

Hotta Y, S Benzer (1970). Genetic dissection of the Drosophila nervous system by means of mosaics. Proc Natl Acad Sci USA 67:1156-1163.

Johnson EC, Pak WL (1986). Electrophysiological study of Drosophila rhodopsin mutants. J Gen Physiol 88:651-673.

Johnson EC, Robinson PR, Lisman JE (1986). Cyclic GMP is involved in the excitation of invertebrate photoreceptors. Nature 324:469-470.

Katan M, Kriz RW, Totty N, Philip R, Meldrum E, Aldape RA, Knopf JL, Parker PJ (1988). Determination of the primary structure of PLC-154 demonstrates diversity of phosphoinositide-specific phospholipase C activities. Cell 54:171-177.

Kubo T, Fukuda K, Mikami A, Maeda A, Takahashi H, Mishina M, Haga T, Haga K, Ichiyama A, Kangawa K, Kojima M, Matsuo H, Hirose T, Numa S (1986). Cloning, sequencing, and expression of complementary DNA encoding the muscarinic acetylcholine receptor. Nature, Lond 323:411-416.

Kyte J, Doolittle RF (1982). A simple method for displaying the hydropathic character of a protein. J Mol Biol 157:105-132.

Lang K, Schmid FX, Fischer G (1987). Catalysis of protein folding by prolyl isomerase. Nature 329:268-270.

Larrivee DC, Conrad SK, Stephenson RS, Pak WL (1981). Mutation that selectively affects rhodopsin concentration in the peripheral photoreceptors of Drosophila melanogaster. J Gen Physiol 78:521-545.

Levy LS, Ganguly R, Ganguly N, Manning JE (1982). The selection, expression, and organization of a set of head-specific genes in Drosophila. Dev Biol 94:451-464.

Matsumoto H, Isono K, Pye Q, Pak WL (1987). Gene encoding cytoskeletal proteins in Drosophila rhabdomeres. Proc Natl Acad Sci USA 84:985-989.

Minke B (1982). Light-induced reduction in excitation efficiency in the trp mutant of Drosophila. J Gen Physiol 79:361-385.

Minke B, Wu C-F, Pak WL (1975). Induction of photoreceptor voltage noise in the dark in Drosophila mutant. Nature 258:84-87.

Montell C, Rubin GM (1988). The Drosophila ninaC locus encodes two photoreceptor cell specific proteins with domains homologous to protein kinases and the myosin heavy domain head. Cell 52:757-772.

Montell C, Rubin GM (1989). Molecular characterization of the Drosophila trp locus: A putative integral membrane protein required for phototransduction. Neuron 2:1313-1323.

Montell C, Jones K, Hafen E, Rubin G (1985). Rescue of the Drosophila phototransduction mutation trp by germline transformation. Science 230:1040-1043.

Montell C, Jones K, Zuker C, Rubin G (1987). A second opsin gene expressed in the ultraviolet sensitive R7 photoreceptor cells of Drosophila melanogaster. J Neurosci 7:1558-1566.

Nathans J, Hogness DS (1983). Isolation, sequence analysis and intron-exon arrangement of the gene encoding bovine rhodopsin. Cell 34:807-814.

Nathans J, Hogness DS (1984). Isolation and nucleotide sequence of the gene encoding human rhodopsin. Proc Natl Acad Sci USA 81:4851-4855.

O'Brien SJ, MacIntyre RJ (1978). Genetics and biochemistry of enzymes and specific proteins of Drosophila In Ashburner M, Wright TRF (eds): "The Genetics and Biology of Drosophila, Vol 2a," New York: Academic Press, pp 396-551.

O'Tousa JE, Baehr W, Martin RL, Hirsh J, Pak WL, Applebury ML (1985). The Drosophila ninaE gene encodes an opsin. Cell 40:839-850.

O'Tousa JE, Leonard DS, Pak WL (1989). Morphological defects in ora^{JK84} photoreceptors caused by mutation in R1-6 opsin gene of Drosophila. J Neurogenet, in press.

Ovchinnikov YA, Abdulaev NG, Feigina MY, Artamonov ID, Zolotarev AS, Miroshnikov AI, Martynov UI, Kustina MB, Kudelin AB, Bogachuk AS (1982). The complete amino acid sequence of visual rhodopsin. Bioorg Khim 8:1011-1014.

Pak WL (1975). Mutations affecting the vision of Drosophila melanogaster. In King RC (ed): "Handbook of Genetics, Vol 3," New York: Plenum Publishing Corp, pp 703-733.

Pak WL (1979). Study of photoreceptor function using Drosophila mutants. In Breakefield, XO (ed): "Neurogenetics: Genetic Approaches to the Nervous System," New York: Elsevier North Holland, pp 67-99.

Pak WL, O'Tousa JE (1988). Molecular analysis of visual pigment genes. Photochemistry and Photobiology 47:877-882.

Pak WL, Grossfield J, White NV (1969). Nonphototactic mutants in a study of vision of Drosophila. Nature 222:351-354.

Pak WL, Grossfield J, Arnold K (1970). Mutants of the visual pathway of Drosophila melanogaster. Nature 227:518-520.

Pak WL, Ostroy SE, Deland MC, Wu CF (1976). Photoreceptor mutant of Drosophila: Is protein involved in intermediate steps of phototransduction? Science 194:956-959.

Pak WL, Conrad SK, Kremer NE, Larrivee DC, Schinz RH, Wong F (1980). Photoreceptor function. In Siddiqi O, Babu P, Hall LM, Hall JC (eds): "Development and Neurobiology of Drosophila," New York: Plenum Publishing Corp, pp 331-346.

Palazzolo MJ, Hyde DR, Raghavan KV, Mecklenburg K, Benzer S, Meyerowitz EM (1987). Isolation and preliminary characterization of 500 non-cross-hybridizing Drosophila "head-specific" cDNA clones by sorting a subtracted λSWAJ cDNA library. In "Molecular Neurobiology of Drosophila Abstr", Cold Spring Harbor, NY: Cold Spring Harbor Laboratories, p 53.

Payne, R (1986). Phototransduction by microvillar photoreceptors of invertebrates: mediation of a visual cascade by inositol trisphosphate. Photobiochem Photobiophys 13:373-397.

Rhee SG, Suh P-G, Ryu S-H, Lee SY (1989). Studies of inositol phospholipid-specific phospholipase C. Science 244:546-550.

Rubin GM, Spradling AC (1982). Genetic transformation of Drosophila with transposable element vectors. Science 218:348-353.

Rushton WAH (1961). The intensity factor in vision. In McElroy WD, Glass HB (eds): "Light and Life," Baltimore: The Hopkins University Press, pp 706-722.

Saibil HR (1982). An ordered membrane-cytoskeleton network in squid photoreceptor microvilli. J Mol Biol 158:435-456.

Scavarda NJ, O'Tousa J, Pak WL (1983). Drosophila locus with gene-dosage effects on rhodopsin. Proc Natl Acad Sci USA 80:4441-4445.

Schaeffer E, Smith D, Mardon G, Quinn W, Zuker C (1989). Isolation and characterization of two new Drosophila protein kinase C genes, including one specifically expressed in photorceptor cells. Cell 57:403-412.

Schinz RH, Lo M-VC, Larrivee DC, Pak WL (1982). Freeze-fracture study of the Drosophila photoreceptor membrane: mutations affecting membrane particle density. J Cell Biol 93:961-969.

Schneuwly S, Shortridge RD, Larrivee DC, Ono T, Ozaki M, Pak WL (1989). Drosophila ninaA gene encodes an eye-specific cyclophilin (cyclosporine A binding protein). Proc Natl Acad Sci USA 86:5390-5394.

Selinger Z, Minke B (1988). Inositol lipid cascade of vision studied in mutant flies. In Watson JD, Wigler M, Feramisco J (eds): "Cold Spring Harbor Symp. Quant. Biol, Vol 53: The Molecular Biology of Signal Transduction," Cold Spring Harbor, New York: Cold Spring Harbor Laboratory, pp 333-341.

Shieh B-H, Stamnes MA, Seavallo S, Harris G, Zuker CS (1989). ninaA, a gene required for visual transduction in Drosophila encodes a homolog of cyclosporine A binding protein. Nature 338:67-70.

Spradling AC, Rubin GM (1982). Transposition of cloned P elements into Drosophila germ line chromosomes. Science 218:341-347.

Stahl ML, Ferenz CR, Kelleher KL, Kriz RW, Knopf JL (1988). Sequence similarity of phospholipase C with the non-catalytic region of src. Nature 332:269-272.

Stephenson RS, O'Tousa J, Scavarda NJ, Randall LL, Pak WL (1983). Drosophila mutants with reduced rhodopsin content. In Cosens DJ, Vince-Price D (eds): "The Biology of Photoreception," Cambridge, England: Cambridge University Press, pp 447-501.

Stryer L (1983). Transducin and the cyclic GMP phosphodiesterase: Amplifier proteins in vision. In "Cold Spring Harbor Symposium on Quantitaive Biology, Volume XLVII," Cold Spring Harbor, NY: Cold Spring Harbor Laboratory.

Stryer L (1986). Cyclic GMP cascade of vision. Ann Rev Neurosci 9:87-119.

Suh, PG, Ryu SH, Moon KH, Suh HW, Rhee SG (1988a). Cloning and sequence of multiple forms of phospholipase C. Cell 54:161-169.

Suh PG, Ryu SH, Moon KH, Suh HW, Rhee SG (1988b). Inositol phospholipid-specific phospholipase C: Complete cDNA and protein sequences and sequence homology to tyrosine kinase-related oncogene products. Proc Natl Acad Sci USA 85:5419-5423.

Szutz EZ, Wood SF, Reid MS, Fein A (1986). Light stimulates the rapid formation of inositol trisphosphate in squid retinas. Biochem J 240:929-932.

Takahashi N, Hayano T, Suzuki M (1989). Peptidyl-prolyl cis-trans isomerase is the cyclosporin A-binding protein cyclophilin. Nature 337:473-475.

Van Vactor D Jr, Krantz DE, Reinke R, Zipursky SL (1988). Analysis of mutants in chaoptin, a photoreceptor cell-specific glycoprotein in Drosophila, reveals its role in cellular morphogenesis. Cell 52:281-290.

Wong F, Schaefer EL, Roop BC, LaMendola JN, Johnson-Seaton D, Shao D (1989). Proper function of the Drosophila trp gene product during pupal development is important for normal visual transduction in the adult. Neuron 3:81-94.

Wu C-F, Pak WL (1978). Light-induced voltage noise in the photoreceptor of Drosophila melanogaster. J Gen Physiol 71:249-268.

Yau KW, Nakatani K (1985). Light-suppressible, cyclic GMP-sensitive conductance in the plasma membrane of a truncated rod outer segment. Nature, Lond 317:252-255.

Yoshioka T, Inoue H, Hotta Y (1985). Absence of phosphatidylinositol phosphodiesterase in the head of a Drosophila visual mutant, norpA (no receptor potential A).

J Biochem 97:1251-1254.

Zuker CS, Cowman AF, Rubin GM (1985). Isolation and structure of a rhodopsin gene from D. melanogaster. Cell 40:851-858.

Zuker CS, Montell C, Jones K, Laverty T, Rubin G (1987). A rhodopsin gene expressed in photoreceptor cell R7 of the Drosophila eye: Homologies with other signal-transducing molecules. J Neurosci 7:1537-1550.

Zuker CS, D Mismer G, Hardy R, Rubin GM (1988). Ectopic expression of a minor Drosophila opsin in the major photoreceptor cell class: Distinguishing the role of primary receptor and cellular context. Cell 55:475-482.

Molecular Biology of the Retina: Basic and Clinically Relevant Studies, pages 33–66

THE MOLECULAR GENETICS OF RETINAL PHOTORECEPTOR PROTEINS INVOLVED IN cGMP METABOLISM[1]

Steven J. Pittler and Wolfgang Baehr

Cullen Eye Institute, Department of Ophthalmology, Baylor College of Medicine, Houston, Texas 77030

INTRODUCTION

3′,5′-Cyclic guanosine monophosphate (cGMP) was identified in mammalian tissue homogenates more than 25 years ago (Ashman et al., 1963). The enzymes responsible for synthesis and hydrolysis of cyclic nucleotides, phosphodiesterases and cyclases, were reported shortly thereafter (for review see Goldberg et al., 1973). Since its discovery, the elucidation of the role of cGMP as an intracellular second messenger has always lagged behind that of cAMP. Up to now, three major regulatory mechanisms have been described for cGMP: cGMP mediates muscular relaxation, it causes efflux of sodium (natriuresis) and vasodilatation upon excitation by atrial natriuretic peptides (ANP), and it interacts directly with Na^+ channels in phototransduction. Excellent reviews of the role of cGMP in various tissues, and of the proteins involved in its metabolism, are available (Goldberg and Haddox, 1977; Tremblay et al., 1988; Beavo, 1988). The best characterized intracellular regulatory mechanism involving cGMP is that of the phototransduction cascade that mediates vision in mammalian rod photoreceptors.

---

[1]This work was supported by a grant from the National Retinitis Pigmentosa Foundation (RP Foundation Fighting Blindness) and a Jules and Doris Stein Professorship from Research to Prevent Blindness, Inc. to W.B. S.J.P. is the recipient of an Individual National Research Service Award from the National Eye Institute.

Phototransduction is defined as the sequence of events that link light-activation of the photoreceptor molecule to production of a neural signal. In both rods and cones, a cascade of biochemical events leads to a transient decrease in Na^+ conductance of the photoreceptor plasma membrane (reviewed in: Stryer, 1986; Pugh and Cobbs, 1986; Chabre and Deterre, 1989). The protein components of the cascade are best characterized in rod photoreceptors which outnumber cones in most mammalian retinas 10-100 to 1. The rod cascade begins when light activates the photoreceptor, rhodopsin (figure 1). Activated rhodopsin (R^*) catalytically interacts with a photoreceptor outer segment G protein, transducin, causing rapid exchange of cytoplasmic GTP with GDP bound to the alpha-subunit (Fung and Stryer, 1980; Fung et al., 1981). The alpha-subunit charged with GTP in turn activates a second amplifier protein (Uchida et al., 1981), cGMP phosphodiesterase (PDE), presumably by direct interaction with an intrinsic inhibitory subunit (Yamazaki et al.,

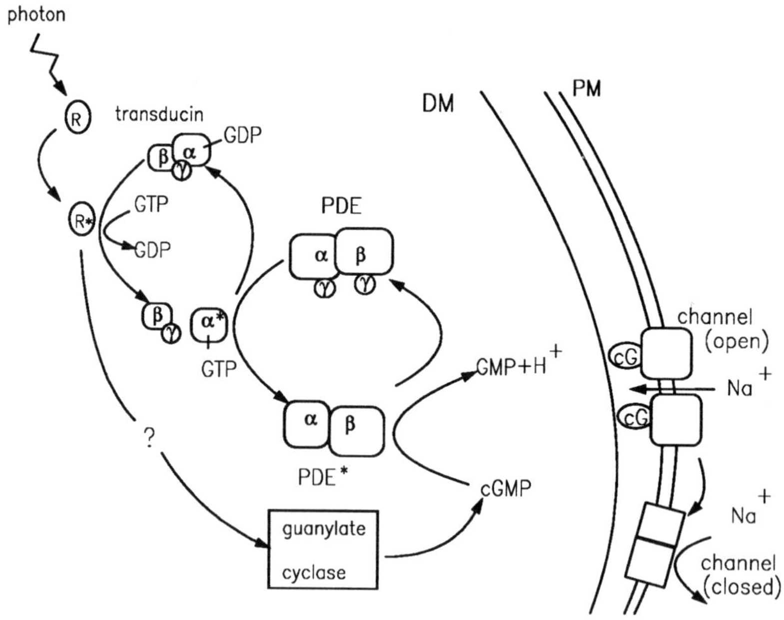

Figure 1. Artist's view of the phototransduction cascade. Shown are the major protein components of the cascade (explanations see text). DM, disk membrane; PM, plasma membrane.

1983). Activated PDE now rapidly hydrolyzes cytoplasmic cGMP which is restored to a basal level in the dark by a guanylate cyclase. The turnover number of activated PDE increases 20-80 fold to 2,000-3,700 moles cGMP/sec (Baehr et al., 1979; Stryer, 1985). Depletion of cGMP in the cytoplasm causes cGMP-dependent Na^+-channels in the outer membrane to close effecting hyperpolarization (Fesenko et al., 1985). The amplification, speed, and sensitivity of the cascade is most remarkable. A single photon may cause hydrolysis of millions of cGMP molecules (Yee and Liebman, 1978) in less than one second.

Cones differ dramatically from rods in several ways, but the biochemistry of phototransduction is thought to be similar (Pugh and Cobbs, 1986). The components of cone phototransduction have been characterized mostly through recombinant DNA and electrophysiological techniques. The genes encoding blue and red/green pigments of human cones were shown to be homologous to rod rhodopsin genes (Nathans et al., 1986). A cone outer segment G protein alpha-subunit was characterized by molecular cloning and by immunocytochemical techniques (Lerea et al., 1986). A cone PDE was recently purified (see below). The cone cGMP-dependent conductance in the plasma membrane was shown to be similar to that in rods (Haynes and Yau, 1985).

The protein components of phototransduction (rhodopsin, the rod outer segment G protein (transducin), cGMP phosphodiesterase, cGMP-gated cation channel), its regulators (rhodopsin kinase, phospho-opsin phosphatase, arrestin), and general mechanisms of amplification and regulation, have been reviewed in great detail elsewhere (Stryer, 1986; Nathans, 1987; Liebman et al., 1987; Hurley, 1987; Falk and Applebury, 1988; Chabre and Deterre, 1989). In this contribution, we will focus on recent progress concerning the molecular biology, particularly the status of cloning and gene characterization, of the two types of photoreceptor proteins involved in synthesis (guanylate cyclase) and hydrolysis (cGMP phosphodiesterases) of cGMP. The emphasis will be on recent advances in the molecular genetics of the cGMP phosphodiesterase.

PHOTORECEPTOR GUANYLATE CYCLASE

The isolation of photoreceptor cyclases has not been achieved, and information about structure is currently not available. Purified guanylate cyclases from other tissues (e.g., bovine lung) exist in two different forms: a single-subunit transmembrane protein and a two-subunit soluble, haem-containing enzyme (briefly reviewed in "News and Views", Newmark, 1989). A membrane-spanning guanylate cyclase has recently been shown (Chinkers et al., 1989; Schulz et al., 1989) to function extracellularly as a receptor for atrial natriuretic peptides (ANPs), and intracellularly, upon excitation by ANP, as a guanylate cyclase producing cGMP. This protein thus represents the first known receptor that directly produces a second messenger upon stimulation by a ligand. The two functional domains of the receptor are separated by a single trans-membrane domain. The catalytic domain is homologous to one of the subunits of the soluble guanylate cyclase from bovine lung, which has recently been cloned and sequenced (Koesling et al., 1988) indicating the existence of a complex and diverse guanylate cyclase gene family.

In bovine photoreceptors, the axoneme-basal apparatus complex of the outer segment has been shown to contain high levels of guanylate cyclase activity (Pannbacker, 1973; Fleischman and Denisevich, 1979). This activity is sensitive to Ca^{++} ions, and cannot be separated from the axonemes by extraction with nonionic detergents, alkali salts, or EDTA, suggesting that the enzyme may be a member of the membrane-spanning cyclases. Recently it was suggested that the enzyme is regulated by a highly cooperative calcium switch that senses changes in the calcium concentration in the submicromolar range, and that the activity is stimulated at low calcium by a peripherally membrane-bound regulatory protein (Koch and Stryer, 1988).

PHOTORECEPTOR cGMP PHOSPHODIESTERASES

Photoreceptor cGMP phosphodiesterases have been characterized in several eucaryotic systems, especially amphibian and mammalian. Tryptic peptide mapping of various vertebrate rod PDEs indicated a high sequence conservation between species (Takemoto et al., 1984). Bovine rod PDE holoenzyme was shown to be a peripherally membrane-bound enzyme consisting of three subunits (Baehr et al., 1979), termed alpha, beta, gamma, with SDS-PAGE mobilities of 88K (alpha), 84K (beta), and 13K (gamma). Apparent mobilities of corresponding subunits of frog PDE are slightly slower (Miki et al., 1975; Whalen and Bitensky, 1989). Alpha- and beta-subunits contain the cGMP-binding and catalytic site(s). It was recently shown that one mole of purified PDE contains 2 moles of tightly bound cGMP (Gillespie and Beavo, 1989), perhaps one on each of the two large subunits. The alpha-subunit contains a domain (figure 6) that is homologous to other eucaryotic cyclic nucleotide phosphodiesterases (Beavo, 1988). The beta-subunit, after separation from PDE holoenzyme in sub-microgram amounts by HPLC, also exhibited cGMP-hydrolyzing activity (Hurwitz et al., 1985) suggesting that it may contain a similar domain. According to peptide patterns obtained by limited proteolysis (Swanson and Applebury, 1983; Baehr, unpublished results), the two large subunits are related indicating a similar amino acid composition. The small gamma-subunit was shown to act as an inhibitor in the dark adapted state (Hurley and Stryer, 1982) constraining the turnover of cGMP by PDE holoenzyme to about 45 moles cGMP/sec. The molecular weight of purified bovine rod PDE holoenzyme was determined to be approximately 170,000 $\pm$ 10,000 by analytical ultracentrifugation (Baehr et al., 1979), and approximately 180,000 in the membrane bound state by target analysis (Hughes and Brand, 1983). The stoichiometric composition of rod PDE was recently suggested to be 1:1:2 (Deterre et al., 1988; Fung and Griswold-Prenner, 1989), with two classes of gamma binding sites (Whalen and Bitensky, 1989). Upon interaction with G-alpha-GTP, rod PDE increases its turnover of cGMP to 3,700 moles cGMP/mole/sec thus amplifying the signal received by rhodopsin (Stryer, 1985). In addition to this indirect light-activation, rod PDE (Miki et al., 1975) may be nonphysiologically activated by limited proteolysis (trypsin) or polycations (e.g., protamine). Activation is thought to be achieved by proteolytic destruction or physical removal of the inhibitory gamma-subunit(s) (Yamazaki et al., 1983; Deterre et al., 1988; Fung and Griswold-Prenner, 1989).

The membrane association of PDE holoenzyme was shown to be dependent on the ionic strength of the buffer in which the ROS membranes are suspended, and independent of the bleaching state of rhodopsin (in contrast to transducin). In moderate ionic strength, PDE is peripherally bound to the disk membrane; in low ionic strength (hypotonic shock), PDE solubilizes reversibly allowing rapid and effective purification of the enzyme (Baehr et al., 1979; Kühn, 1980). Activation of bovine and frog rod PDE by trypsin results in irreversible release of PDE from the membrane surface (Wensel and Stryer, 1986; Whalen and Bitensky, 1989) indicating the loss of a membrane binding domain by proteolysis.

Cone PDE has not been investigated as thoroughly as rod PDE because of the difficulty in obtaining cone photoreceptor components. Bovine cone PDE, however, was separated from rod PDE by HPLC and purified. It was suggested to consist of at least one large subunit, alpha', whose mobility is slightly slower (95K) than that of rod PDE, and several small peptides one of which displays properties very similar to the rod inhibitor (Hurwitz et al., 1985; Gillespie and Beavo, 1988). The molecular weight of the native cone PDE was found to be roughly comparable to that of rod PDE indicating a dimeric structure of the core enzyme.

An extracellular cGMP phosphodiesterase has been characterized in the interphotoreceptor matrix (Barbehenn et al., 1985). This PDE has subunit sizes of 47K and 45K with an apparent native molecular weight of 350,000 as determined by gel filtration. Its kinetic properties are essentially identical to those observed for the rod PDE, however, its function and cellular origin are unknown. A cGMP phosphodiesterase activity was also reported in the outer plexiform layer of the retina (Pannbacker and Lovett, 1977), but no further characterization of this enzyme has been achieved.

Characterization and sequence analysis of rod PDE alpha-subunit cDNA clones of bovine, human and mouse.

Preliminary analysis of cDNA clones encoding bovine rod PDE large subunits were reported in the laboratory of R.L. Davis (Pittler et al., 1987). However, the identity of these clones as alpha or beta was not determined. A primary sequence of the alpha-subunit was reported from the laboratory of Y. Ovchinnikov (Ovchinnikov et al., 1987). Evidence supporting the identity of nine overlapping alpha-subunit cDNA clones was suggested by the isolation and sequence analysis of two alpha-specific peptides. Further support came from the sequence determination of the first 10 amino acids at the N-terminus. More recently an antipeptide antibody has been produced against the N-terminal sequence

(D. Takemoto, personal communication). This antibody recognizes only the alpha-subunit by western blot analysis confirming the identity of the alpha-subunit cDNA clones. Recently, our laboratory described the isolation of full length cDNA clones for the bovine, human (Pittler et al., 1990) and mouse (Baehr and Pittler, 1990) PDE alpha-subunit (see below). Maps of the cDNA clones that were used to deduce the primary structures are shown in Fig 2.

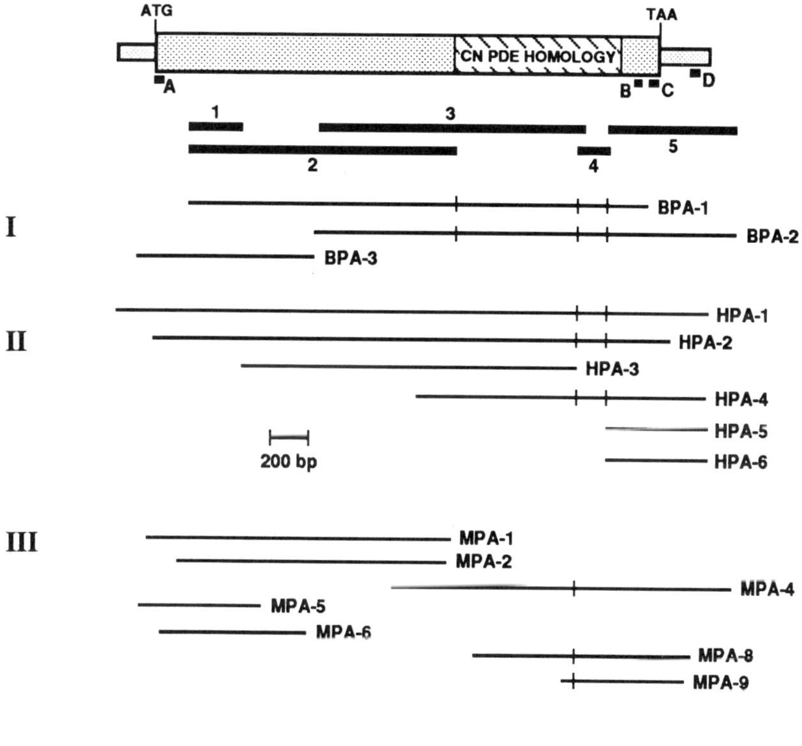

Figure 2. cDNA clones of bovine (I), human (II) and mouse (III) rod photoreceptor PDE alpha-subunit. The large box starting with ATG and ending with TAG depicts the coding sequence. Boxes labelled A - D indicate the positions of oligonucleotide probes, larger boxes labelled 1 - 5 indicate the extent of cDNA probes. HPA-1 is a full-length copy cDNA clone in lambda-gt10. All other clones are plasmids (pUC or bluescript) that end or start at either one of two internal EcoRI sites (E).

To isolate bovine cDNA clones, oligonucleotides were designed according to the sequence of tryptic peptides (Pittler et al., 1990). Peptide sequences obtained from the alpha-subunit are underlined in figure 4. A screen of a lambda-gt10 retinal cDNA library (Nathans and Hogness, 1983) yielded six cDNA clones that contain alpha-subunit sequence. Three of the clones designated BPA-1, BPA-2, and BPA-3 overlap to contain the entire open reading frame (figure 2). There are 14 nucleotide differences between the sequence presented in figure 3 and that reported (Ovchinnikov et al., 1987), of which three result in amino acid changes. Those changes are at positions 194 (V to A), 424 (T to A), and 675 (F to C). Heterogeneity observed in three other cDNA clones is presently under further analysis. It is possible that minor isozymes of the retinal PDE alpha-subunit are encoded by these cDNAs.

The human PDE alpha-subunit clones (figure 2) were isolated from a lambda-gt10 library (J. Nathans, Johns Hopkins Medical Center, Baltimore, MD) with portions of bovine cDNA clones and bovine cDNA derived oligonucleotides as probes (Pittler et al., 1990). Several cDNA clones were obtained including one full length clone designated HPA-1 (figure 2). The truncated cDNA clones did not show sequences that deviated from the HPA-1 sequence. A mouse PDE alpha-subunit clone (MPA-9, figure 2) was isolated from a lambda-gt11 mouse retina library (J. McGinnis, UCLA, Los Angeles, CA) using a bovine cDNA probe. The mouse fragment was then used as a probe to screen a lambda-ZapII cDNA library resulting in the recovery of several overlapping C-terminal cDNAs (Baehr and Pittler, 1990). N-terminal cDNA clones were isolated with a human N-terminal EcoRI/BamHI fragment of HPA-1. The DNA sequences encoding the PDE alpha-subunits of bovine, human and mouse were found to be highly homologous (≧92%, figure 3).

```
                    +1
bpa   TGCAATTCCCAGCCATGGGCGAGGTGACGGCAGAGGAAGTAGAGAAGTTT   36
hpa     GG                        A        G  G        C
mpa   A GG G  TTCA       T        A        G     A     C
m                    M  G  E  V  T  A  E  E  V  E  K  F

bpa   CTGGACTCAAATGTCAGCTTTGCCAAACAGTACTACAACCTGCGCTACCG   86
hpa           G   ATG                          C A
mpa        T      ATG              A     T     T A
m      L  D  S  N  I  G  F  A  K  Q  Y  Y  N  L  H  Y  R
```

```
bpa  GGCCAAGGTCATCTCAGACCTGCTGGGACCCAGGGAGGCGGCCGTGGACT
hpa         C       C     C  T  GG   A      T
mpa    GG                 C  C  GG   A      A  T
m      G  K  V  I  S  D  L  L  G  A  K  E  A  A  V  D    45

bpa  TCAGCAACTACCATGCGCTGAACAGCGTGGAAGAGAGTGAAATTATCTTC  186
hpa               CT C C  G    A    G     C     C     T
mpa               C ATG          A  G        G  C     T
m    F  S  N  Y  H  D  V  N  S  V  E  E  S  E  I  I  F

bpa  GACCTCCTGCGGACTTTCAGGACAACCTGCAGGCCGAGAAGTGCGTCTT
hpa    T                    G  TT A   A A     A   A
mpa                 G       G   T A     T     A   ACA
m     D  L  L  R  D  V  Q  E  N  L  Q  A  E  K  C  T  F

bpa  CAATGTCATGAAGAAGTTGTGCTTCCTGCTGCAGGCCGACCGCATGAGCC  286
hpa                  C          C        A
mpa                  C C        C    G   T     AG
m      N  V  M  K  K  L  C  F  L  L  R  A  D  R  V  S    95

bpa  TATTCATGTACAGGGCCCGGAACGGCATCGCAGAGCTGGCCACCCGGCTC
hpa   G         C  A    C  T     GC              A    T
mpa   G            A    C           C           TA
m    L  F  M  Y  R  T  R  N  G  I  A  E  L  A  T  R  L

bpa  TTCAACGTCCACAAGGATGCTGTGCTCGAGGAGTGCCTGGTGGCGCCTGA  386
hpa       T                 C        C         AT   C
mpa       T                    A     C   T     AT   C
m     F  N  V  H  K  D  A  V  L  E  D  C  L  V  M  P  D

bpa  CTCGGAGATCGTGTTCCCCCTGGACATGGGAGTGGTGGGCCATGTTGCGC
hpa   CAA        C     TT          CA C           C  A
mpa     C     T  C     T           T  C        C  C  A
m      S  E  I  V  F  P  L  D  M  G  V  V  G  H  V  A    145

bpa  TCTCTAAAAAGATCGTCAACGTCCCCAACACAGAGGAGGATGAACATTTC  486
hpa  A      G     T CT                          G
mpa  A   C        T C   T              A        G
m    H  S  K  K  I  A  N  V  P  N  T  E  E  D  E  H  F

bpa  TGTGACTTTGTGGACACCCTCACAGAGTACCAGACTAAGAACATCTTGGC
hpa                  T             A    C
mpa          C       AT        A  T     C         C
m     C  D  F  V  D  N  L  T  E  Y  Q  T  K  N  I  L  A

bpa  TTCTCCCATAATGAATGGGAAGGATGTGGTGGCCATAATCATGGTTGTGA  586
hpa     C                                        C
mpa     C     C                    A             C
m      S  P  I  M  N  G  K  D  V  V  A  I  I  M  A  V    195

bpa  ATAAAGTGGATGGGCCCCACTTCACCGAGAATGATGAAGAGATTCTTCTC
hpa               AT           A   GA
mpa       A A    AA            A   GA
m    N  K  I  D  E  P  H  F  T  K  R  D  E  E  I  L  L

bpa  AAGTACCTCAATTTTGCAAATCTAATCATGAAGGTGTTCCACCTGAGTTA  686
hpa                                   TG  A
mpa             C    TG  C  G           A           C
m     K  Y  L  N  F  V  N  L  I  M  K  V  F  H  L  S  Y
```

(continued)

```
bpa   CCTGCACAACTGTGAGACTCGGCGTGGCCAGATACTGTTGTGGTCTGGGA
hpa                  A     A               C
mpa                        T  C          T  C
m       L  H  N  C  E  T  R  R  G  Q  I  L  L  W  S  G    245

bpa   GCAAAGTCTTTGAAGAGCTTACAGACATTGAGAGGCAGTTCCACAAAGCC  786
hpa                   A     G     C  AC A
mpa       G        G     C  G  T  A                 G
m     S  K  V  F  E  E  L  T  D  I  E  R  Q  F  H  K  A

bpa   CTGTACACAGTCCGCGCCTTCCTCAACTGTGACAGATACTCTGTGGGGCT
hpa                 T  T                             T
mpa           G     G  T                       C  A  A
m      L  Y  T  V  R  A  F  L  N  C  D  R  Y  S  V  G  L

bpa   CTTAGACATGACCAAACAGAAGGAATTTTTTGATGTGTGGCCAGTCCTGA  886
hpa                  G                          G  T
mpa                                                T
m       L  D  M  T  K  Q  K  E  F  F  D  V  W  P  V  L    295

bpa   TGGGGGAGGCTCCACCCTACGCTGGTCCCAGGACTCCGGATGGAAGGGAA
hpa       T  A T      T   T                         A
mpa       C         G T   T                A  C
m     M  G  E  A  P  A  Y  S  G  P  R  T  P  D  G  R  E

bpa   ATCAACTTTTACAAGGTCATTGACTACATCCTGCATGGCAAAGAAGACAT  986
hpa     T                                         G
mpa     T     C                          C
m      I  N  F  Y  K  V  I  D  Y  I  L  H  G  K  E  D  I

bpa   CAAAGTCATCCCGAATCCTCCTCCAGACCACTGGGCTTTAGTGAGCGGTC
hpa                     A     T     T           A
mpa                  C  A  CG T            C       T
m       K  V  I  P  N  P  P  A  D  H  W  A  L  V  S  G    345

bpa   TCCCCACTTATGTTGCCCAAAATGGCCTGATTTGCAACATCATGAACGCG 1086
hpa                      G                          T
mpa    A   C    C  G  T        T     C     T  A
m     L  P  P  Y  V  A  Q  N  G  L  I  C  N  I  M  N  A

bpa   CCTTCAGAGGACTTTTTTGCATTCCAGAAAGAGCCTCTGGATGAGTCTGG
hpa                          T        A
mpa      G               A
m      P  A  E  D  F  F  E  F  Q  K  E  P  L  D  E  S  G

bpa   ATGGATGATTAAAAATGTCCTTTCTATGCCAATTGTGAACAAGAAGGAGG 1186
hpa                     G     A     G                 A
mpa   G                 A  C  C     C  C  C           A
m       W  M  I  K  N  V  L  S  M  P  I  V  N  K  K  E    395

bpa   AAATTGTTGGGGTGGCCACGTTTTAAAATCGCAAAGATGGGAAACCCTTT
hpa             A        A     C     T           G
mpa    G  C  C  C        A     C  C              G     C
m     E  I  V  G  V  A  T  F  Y  N  R  K  D  G  K  P  F

bpa   GATGAAATGGATGAGACCCTCATGGAGTCTTTGACCCAGTTCCTGGGCTG 1286
hpa                  G                 T  A  T
mpa     C  T     C                       T     T     A
m      D  D  M  D  E  T  L  M  E  S  L  T  Q  F  L  G  W
```

```
bpa  GTCTGTCTTAAATCCTGACACTTATGAGTTGATGAACAAACTTGAAAACA
hpa                       C       CA     T           T
mpa     A        C        C  C    CC        G  C  G
m      S  V  L  N  P  D  T  Y  E  S  M  N  K  L  E  N     445

bpa  GGAAGGATATTTTCCAAGACATGGTGAAATACCACGTGAAGTGTGACAAT 1386
hpa                  G     A  A     T  T
mpa           C     G     C        T              T  C
m    R  K  D  I  F  Q  D  I  V  K  Y  H  V  K  C  D  N

bpa  GAAGAAATCCAGACAATCTTGAAAACCAGAGAGGTGTATGGGAAGGAGCC
hpa          T    A
mpa              AG                       C  C  A
m     E  E  I  Q  K  I  L  K  T  R  E  V  Y  G  K  E  P

bpa  GTGGGAATGCGAGGAAGAAGAACTGGCTGAGATCCTGCAAGGAGAGCTGC 1486
hpa  A     G  T        G  G                   CG
mpa                 G  G  G                  A    A  T
m      W  E  C  E  E  E  E  L  A  E  I  L  Q  R  E  L     495

bpa  CAGATGCAGACAAATATGAAATTAATAAATTCCACTTCAGCGACCTGCCC
hpa            T     C              T        T   T A
mpa      C  G GTC    C     C  C  G                    A
m    P  D  A  E  S  Y  E  I  N  K  F  H  F  S  D  L  P

bpa  CTGACGGAACTGGAGCTGGTGAAATGTGGAATTCAGATGTACTATGAGCT 1586
hpa    A  A              A           A        T
mpa    C     G              G  C  C  C           C
m     L  T  E  L  E  L  V  K  C  G  I  Q  M  Y  Y  E  L

bpa  CAAAGTGGTGGATAAATTTCACATTCCTCAGGAGGCCCTGGTGCGGTTCA
hpa                             A  A
mpa    G    TG   C  G  C     C  G  A              C
m      R  V  W  D  K  F  H  I  P  Q  E  A  L  V  R  F     545

bpa  TGTATTCCCTGAGTAAGGGATACCGCAGGATCACCTACCACAACTGGCGG 1686
hpa      C              C       A
mpa         G  A  C  A  C     G  A     T
m    M  Y  S  L  S  K  G  Y  R  R  I  T  Y  H  N  W  R

bpa  CACCGCTTCAACGTGGGGCAGACCATGTTCTCCTTGCTGGTGACCGGAAA
hpa                                   C          G
mpa         T                                         A
m     H  G  F  N  V  G  Q  T  M  F  S  L  L  V  T  G  K

bpa  GCTGAACCGATACTTCACAGACCTGGAGGCCTTGGCCATGGTCACCGCCG 1786
hpa           C        G     A                    T  T
mpa        A  G        T  T  A                    T  T
m      L  K  R  Y  F  T  D  L  E  A  L  A  M  V  T  A     595

bpa  CCTTCTGCCATGACATTGACCACAGAGGCACTAACAATCTCTACCAGATG
hpa   T                             C  T  C
mpa         T        C              G     C
m    A  F  C  H  D  I  D  H  R  G  T  N  N  L  Y  Q  M

bpa  AAATCCCAGAACCCACTGGCCAAGCTCCATGGGTCCTCCATCTTGGAAAG 1886
hpa                                        T
mpa       A        C
m     K  S  Q  N  P  L  A  K  L  H  G  S  S  I  L  E  R
```

(*continued*)

```
bpa   ACACCACTTGGAGTTTGGTAAAACACTCTTGAGAGATGAGAGCCTAAATA
hpa                     C        GC C     C        G
mpa   G  T  T           C         C                G
m       H  H  L  E  F  G  K  T  L  L  R  D  E  S  L  N     645

bpa   TCTTTCAGAACCTCAATCGCAGGCAGCATGAGCATGCAATCCACATGATG 1986
hpa          A           TC A              C
mpa       C        G      C             C  G
m     I  F  Q  N  L  N  R  R  Q  H  E  H  A  I  H  M  M

bpa   GACATTGCGATCATTGCTACAGACCTTGCCTTGTATTTCAAGAAAAGGAC
hpa           A        C        C   C             G
mpa        C           C
m      D  I  A  I  I  A  T  D  L  A  L  Y  F  K  K  R  T

bpa   GATGTTCCAAAAGATCGTGGATCAGTCTAAGACATACGAGACTCAGCAGG 2086
hpa                                       T    G G A
mpa   C        G     T           A        T    G ACC
m       M  F  Q  K  I  V  D  Q  S  K  T  Y  E  S  T  Q     695

bpa   AGTGGACGCAGTACATGATGCTGGATCAGACACGGAAGGAAATTGTCATG
hpa          A                 G                 C  T
mpa          C                 G                    G
m     E  W  T  Q  Y  M  M  L  E  Q  T  R  K  E  I  V  M

bpa   GCCATGATGATGACCGCCTGTGATCTCTCAGCCATCACCAAGCCTTGGGA 2186
hpa                                            A  C
mpa                                            A  C
m      A  M  M  M  T  A  C  D  L  S  A  I  T  K  P  W  E

bpa   GGTGCAGAGCAAGGTGGCTCTGCTGGTTGCTGCTGAATTCTGGGAACAAG
hpa      CAG    C    A           G
mpa      A                       G                 G
m       V  Q  S  K  V  A  L  L  V  A  A  E  F  W  E  Q     745

bpa   GTGACCTGGAGCGCACGGTGCTGCAACAGAATCCCATTCCCATGATGGAC 2286
hpa
mpa                   A        G
m     G  D  L  E  R  T  V  L  Q  Q  N  P  I  P  M  M  D

bpa   AGAAACAAGGCGGATGAACTCCCTAAGCTTCAGGTCGGCTTCATCGACTT
hpa           A  A                    A           T
mpa                    G     C        A
m      R  N  K  A  D  E  L  P  K  L  Q  V  G  F  I  D  F

bpa   TGTTTGCACCTTTGTCTACAAGGAATTCTCCCGTTTCCACGAGGAGATCA 2386
hpa               C
mpa      G     T        T     G        A  T  T        T
m       V  C  T  F  V  Y  K  E  F  S  R  F  H  E  E  I     795

bpa   CTCCCATGCTGGATGGGATTACCAACAACCGCAAGGAGTGGAAAGCGCTT
hpa    C  A   T    C     C        T              G
mpa    A                 C  T              A     G     G
m     T  P  M  L  D  G  I  T  N  N  R  K  E  W  K  A  L

bpa   GCTGATGAGTATGAGACCAAGATGAAGGGGCTGGAGGAGGAGAAGCAGAA 2486
hpa              C  TG            T  A       A
mpa              C  AG            CC
m      A  D  E  Y  E  A  K  M  K  A  L  E  E  E  K  Q  K
```

```
bpa     ACAGCAGGCAGCCAACCAAGCAGCCGCAGGAAGTCAACACGGCGGAAAGC
hpa            T G     GTC              A   G CG  G     C
mpa     G              G     T  TT C  G AC  G CA  A  G  C
m         Q  Q  A  A  K  Q  A  A  S  G  N  Q  P  G  G  N    845

bpa     AACCTGGGGGCGGGCCTGCATCCAAGTCTTGCTGCGTCCAATAGCTGACT 2580
hpa      G  CA    T CAA  A          C      A    G  A ACCAC
mpa     C  TCCA   T CA        T     C  T   A    G    A  GC
        P  L  Q  G  A  P  A  S  K  S  C  C  I  Q  *         859
```

Figure 3. cDNA sequences of cGMP phosphodiesterase alpha-subunit of bovine (bpa), human (hpa) and mouse (mpa). Translation start and stop codons of the bpa sequence are shown in bold letters. The amino acid sequence of the predicted mouse alpha-subunit (m) is shown below the nucleotide sequences.

The rod PDE alpha-subunit: An intergenic comparison.

The alpha-subunit cDNA sequences of mouse, bovine and human predict a polypeptide of 859 residues (figure 4) with a molecular weight of approximately 100,000, considerably larger than that predicted by its SDS-PAGE mobility. No leader peptide consensus sequence was detected in the predicted amino acid sequences, consistent with the existence of the alpha-subunit as an unsecreted, soluble protein. The only known posttranslational modifications are the loss of the first residue (Met), acetylation of the second residue (Gly) which represents the N-terminus of the mature peptide (Ovchinnikov et al., 1987), carboxymethylation (Swanson and Applebury, 1983) and lipidation of Cys^{856} (see below). Glycosylation of PDE subunits has not been observed, although an N-V-S consensus sequence as a possible site for attachment of carbohydrate was identified near the N-terminus in the bovine sequence. This site, however, is not conserved in human and mouse PDE alpha-subunits.

```
 1      MGEVTAEEVEKFLDSNIGFAKQYYNLHYRGKVISDLLGAKEAAVDFSNYH  m
                                     A L                    h
                        VS        R  A        PR            b

51      DVNSVEESEIIFDLLRDVQENLQAEKCTFNVMKKLCFLLRADRVSLFMYR  m
        SPS M            F     T   I           Q   M        h
        AL               F D       V           Q   M        b
```

(continued)

```
101      TRNGIAELATRLFNVHKDAVLEDCLVMPDSEIVFPLDMGVVGHVAHSKKI  m
                                      Q         I            h
         A                     E   A                  L      b

151      ANVPNTEEDEHFCDFVDNLTEYQTKNILASPIMNGKDVVAIIMAVNKIDE  m
                          I    K                        V G  h
         V                T                     V       V G  b

201      PHFTKRDEEILLKYLNFVNLIMKVFHLSYLHNCETRRGQILLWSGSKVFE  m
         S                A     WY                           h
             EN           A                                  b

251      ELTDIERQFHKALYTVRAFLNCDRYSVGLLDMTKQKEFFDVWPVLMGEAP  m
                                                         V   h
                                                             b

301      AYSGPRTPDGREINFYKVIDYILHGKEDIKVIPNPPADHWALVSGLPPYV  m
         P                                   P          T    h
         P A                                 P          T    b

351      AQNGLICNIMNAPAEDFFEFQKEPLDESGWMIKNVLSMPIVNKKEEIVGV  m
                      S    A                                 h
                      S    A                                 b

401      ATFYNRKDGKPFDDMDETLMESLTQFLGWSVLNPDTYESMNKLENRKDIF  m
                      E                                      h
                      E                        L             b

451      QDIVKYHVKCDNEEIQKILKTREVYGKEPWECEEEELAEILQRELPDAES  m
                                                   A     DK  h
           M             T                         G     DK  b

501      YEINKFHFSDLPLTELELVKCGIQMYYELRVWDKFHIPQEALVRFMYSLS  m
                                      K V                    h
                                      K V                    b

551      KGYRRITYHNWRHGFNVGQTMFSLLVTGKLKRYFTDLEALAMVTAAFCHD  m
             K                                               h
                                                             b

601      IDHRGTNNLYQMKSQNPLAKLHGSSILERHHLEFGKTLLRDESLNIFQNL  m
                                                             h
                                                             b

651      NRRQHEHAIHMMDIAIIATDLALYFKKRTMFQKIVDQSKTYESTQEWTQY  m
                                                    E        h
                                                   TQ        b

701      MMLEQTRKEIVMAMMMTACDLSAITKPWEVQSKVALLVAAEFWEQGDLER  m
                                         Q                   h
            D                                                b
```

```
751   TVLQQNPIPMMDRNKADELPKLQVGFIDFVCTFVYKEFSRFHEEITPMLD  m
                                                          h
                                                          b

801   GITNNRKEWKALADEYEAKMKALEEEKQKQQAAKQAASGNQPGGNPLQGA  m
                      D    VQ  K     S  S  A       QPR    h
                       T   G           N   A S H  KQPG G  b

851   PASKSCCIQ  m
      TT         h
             V   b
```

Figure 4. Intergenic comparison of the amino acid sequences of the PDE alpha-subunit of mouse (m), human (h), and bovine (b). Tryptic peptides used to construct oligonucleotide primers are underlined.

The alpha-subunits are acidic polypeptides with an excess of uniformly distributed negatively charged residues. The amino acid sequences are 93.5 to 94.4% identical in the three species. The majority of deviations appear to be concentrated in the first 240 amino acids at the N-terminus, and the last 43 amino acids at the C-terminus of the polypeptide chain. A CAAX (C = cysteine; A = aliphatic, X = any residue) consensus sequence (Magee and Hanley, 1988) at the C-terminus that appears to signal a multistep posttranlational processing (Hancock et al., 1989), however, is conserved in the bovine, human and mouse C-termini. It has been shown that all mammalian *ras* proteins containing the CAAX motif are isoprenylated (Hancock et al., 1989) and carboxymethylated (Clarke et al., 1988), coincident with removal of the last three amino acids (Gutierrez et al., 1989). In the bovine rod PDE alpha-subunit, similar processing has been shown but no identification of the lipid could be made (B. K.-K. Fung, personal communication). In the bovine rod transducin gamma-subunit, four C-terminal residues conforming to the CAAX sequence were apparently removed during processing (Hurley, 1987). Using mammalian cells expressing *ras* proteins it was shown that the inhibition of C-terminal processing abolishes membrane binding and leads to a loss of transforming ability (Hancock et al., 1989). Addition of a lipid side chain to the PDE alpha-subunit and concomitant increase in hydrophobicity may anchor the PDE in the phospholipid bilayer, consistent with the observation that limited proteolysis removes disk membrane binding ability (Wensel and Stryer, 1986; Whalen and

Bitensky, 1989). Further characterization of the role of this processing may uncover important aspects of PDE regulation.

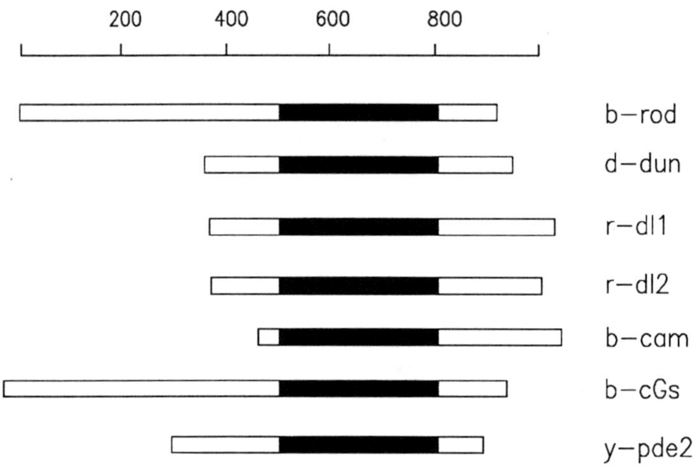

Figure 5. Schematic diagram of various eucaryotic cyclic nucleotide phosphodiesterases. The open boxes are proportional to the lengths of the polypeptide chains predicted from cDNA sequencing and peptide sequence analysis, and the filled boxes denote the approximate region of homology. The scale is in amino acid residues. b-rod, bovine retina cGMP phosphodiesterase; d-dun, Drosophila *dunce* cAMP PDE; r-dl1 and r-dl2, rat brain *dunce*-like PDEs; b-cam, bovine brain Ca^{++}/calmodulin-dependent PDE; b-cGS, bovine brain cGMP-stimulated phosphodiesterase; y-pde2, PDE2 of yeast.

and a Dictyostelium PDE, the other the low-K_m yeast PDE (PDE2, Sass et al., 1986), the rod cGMP PDE, the Drosophila *dunce* cAMP PDE (Chen et al., 1986), two recently identified *dunce*-like PDEs in rat brain (Colicelli et al., 1989; Davis et al., 1989), cGMP-stimulated and Ca^{++}/Calmodulin-stimulated cNMP phosphodiesterases (Charbonneau et al., 1986). As shown in figure 5, these PDEs share an approximately 300 residue homologous domain containing about 25 identities from yeast to man. This domain is thought to contain at least part of the active site of the enzyme.

The region best conserved between the three mammalian photoreceptor PDE alpha-subunits (pos. 532- 826, figure 3) encompasses a domain that is also conserved between several eucaryotic cyclic nucleotide phosphodiesterases (Charbonneau et al., 1986). Based on peptide mapping, sequence similarities were also suggested to exist among diverse mammalian PDEs (Takemoto et al., 1982). To date, at least two major classes of PDEs have been identified based on comparison of biochemical properties and available sequence information (Beavo, 1988). One class contains the higher K_m yeast PDE

```
511  L P L T E L E L V K C G I Q M Y Y E L K V V D K F   b-rod
203  R P L T C V A Y T - - - I F Q S R E L L T S - - L   d-dun
141  R S L S C I M Y T - - - I F Q E R D L L K K - - F   r-dl1
143  R P L T C I M Y A - - - I F Q E R D L L K T - - F   r-dl2
                                                 K I   b-cam
                                                 K I   b-cGS
216  L S L S T Q E L I W C G F T L I K K L S K D A K V   y-pde2

536  H I P Q E A L V R F M Y S L S K G Y R R - I T Y H   b-rod
223  M I P P K T F L N F M S T L E D H Y V K D N P F H   d-dun
161  H I P V D T M M M Y M L T L E D H Y H A D V A Y H   r-dl1
163  K I S S D T F V T Y M M T L E D H Y H S D V A Y H   r-dl2
     P V S C - L I A F A E A L E V G Y X K Y K N P Y H   b-cam
     D C P T - L A R F C - - L M V K K G Y R D L P Y H   b-cGS
241  L I A D N K L L L L L F T L E S S Y H Q V N K F H   y-pde2

560  N W R H G F N V G Q T M F S L L V T G K L K R Y F   b-rod
248  N S L H A A D V T Q S T N V L L N T P A L A G V F   d-dun
186  N S L H A A D V L Q S T H V L L A T P A L D A V F   r-dl1
188  N S L H A A D V A Q S T H V L L S T P A L D A V F   r-dl2
     N L I H A A D V T Q T V H Y I M L G T G I M H W L   b-cam
     N W M H A F S V S H F C Y L L Y K N L E L T N Y L   b-cGS
266  N F R H A I D V M Q A T W R L C - T Y L L K D N P   y-pde2

585  T D L E A L A M V T A A F C H D I D H R G T N N L   b-rod
274  T P L E V G G A L F A A C I H D V D H P G L T N Q   d-dun
211  T D L E I L A A L F A A A I H D V D H P G V S N Q   r-dl1
213  T D L E I L A A I F A A A I H D V D H P G V S N Q   r-dl2
     T E L E I L A M V F A A A I H D Y E H T G T T N N   b-cam
     E D M E I F A L F I S C M C H D L D H R G T N N S   b-cGS
290  V Q T L L L C M - - A A I G H D V G H P G T N N Q   y-pde2

610  Y Q M K S Q N P L A K L H G - - S S I L E R H H L E   b-rod
298  F L V N S S S E L A L M Y - N D E S V L E N H H L A   d-dun
236  F L I N T N S E L A L M Y - N D E S Y L E N H H L A   r-dl1
238  F L I N T N S E L A L M Y - N D E S V L E N H H L A   r-dl2
     F H I Q X R S D V A I L Y - N D R S V L E N H H V S   b-cam
     F Q V A S K S V L A A L Y S S E G S V M E R H H F A   b-cGS
313  L L C N C E S E V A Q N F K N V - S I L E N F H R E   y-pde2

634  F G K T L L R D E S L N I F Q N L N R R Q H E H A   b-rod
323  V A F K L L Q N Q G C D I F C N M Q K K Q R Q T L   d-dun
261  V G F K L L Q E E N C D I F Q N L S K R Q R Q S L   r-dl1
263  V G F K L L Q E E H C D I F Q N L T K K Q R Q T L   r-dl2
     A A Y R L M Q E E E M N V L I N L S K - D D W R D   b-cam
     Q A I A I L N T H G C N I F D H F S R K D Y Q R M   b-cGS
338  L F Q Q L L S E H W P L K L S I S K K K F D F I S   y-pde2
```

(continued)

```
659  I H M M - - D I A I I A T D L A L Y C K K R T M F   b-rod
348  R K M V I - D I V L S T D M S K H M S L L A D - -   d-dun
286  R K M V I - D M V L A T D M S K H M T L L A D - -   r-dl1
288  R K M V I - D M V L A T D M S K H M S L L A D - -   r-dl2
     L R W L V I E M V L S T D M S G H F Q Q I K N - -   b-cam
     - L D L M R D I I L A T D L A H H L R I F K D L Q   b-cGS
363  E A I L A T D M A L H S Q Y E D R L M H E - - - -   y-pde2

682  Q K I V D Q S K T - Y E T Q Q E W T Q Y - M M L D   b-rod
370  - - - - - - L K T M V E T K K V A G S G V L L L D   d-dun
308  - - - - - - L K T M V E T K K V T S S G V L L L D   r-dl1
310  - - - - - - L K T M V E T K K V T S S G V L L L D   r-dl2
     - - - - - - - - - - - - - - - - - I R N S L Q Q P   b-cam
     - - - - - K M A E V G Y D R T N K Q H H S L L L C   b-cGS
     - - - - - - - - - - - - - - - - - - - - - - - - -   y-pde2

705  Q T R K E I V M A M M M T - A C D L S A I T K P W   b-rod
389  N Y T D R I Q V L E N L V H C A D L S N P T K P L   d-dun
327  N Y S D R I Q V L R N M V H C A D L S N P T K P L   r-dl1
329  N Y T D R I Q V L R N M V H C A D L S N P T K S L   r-dl2
     E G L D K A K T M S L I L H A A D I S H P A K S W   b-cam
     - - - - - - - - - L L M T - S C D L S D Q T K K -   b-cGS
384  N P M K Q I T L I S L I I K A A D I S N V T R T L   y-pde2

729  E V Q S K - V A L L V A A E F W E Q G D L E R T V   b-rod
414  P L Y K R W V A L L M - E E F F L Q G D K E R E S   d-dun
352  E L Y R Q W T D R I M - A E F F Q Q G D R E R E R   r-dl1
354  E L Y R Q W T D R I M - E E F F Q Q G D K E R E R   r-dl2
     X K L H R W T M A L M - E E F F L Q G D K E A E L   b-cam
     - - - - - - - - - - - - - E F F S Q G D L E K A M   b-cGS
409  S I S A R W A - Y L I T L E F N D C A L L E T F H   y-pde2

753  L Q Q N P I P M M - - D R N K A D E L P K L Q V G   b-rod
438  G M D I - S P M C - - D R H N A T I E - K S Q V G   d-dun
376  G M E I - S P M C - - D K H T A S V E - K S Q V G   r-dl1
378  G M E I - S P M C - - D K H T A S V E - K S Q V G   r-dl2
     G L P F - S P L C - - D R K S T M V A - Q S Q I G   b-cam
     G N R - - - P M E M M D R E K A Y - I P E L Q I S   b-cGS
433  K A H R P E Q D C F G D S Y K N V D S P K E D L E   y-pde2

776  F I D F V C T F V Y K E F - S R F H E E I T P M L   b-rod
459  F I D Y I V H P L W E T W A S L V H P D A Q D I L   d-dun
397  F I D Y I V H P L W E T W A D L V H P D A Q D I L   r-dl1
399  F I D Y I V H P L W E T W A D L V Q P D A Q D I L   r-dl2
     F I D F I V E P                                     b-cam
     F M E H I A M P I Y K L L Q D L F P K A A E L Y E   b-cGS
458  S I Q N I L V N V T D P D D I I K D H P H I P       y-pde2
```

Figure 6. Homologous domains in various cNMP phosphodiesterases. Single letters symbolize amino acid residues (X = unidentified). Gaps (-) were introduced to optimize alignments. Boxes depict residues which are identical (D=E; R=K; I=L=V=M) in seven sequences shown. The abbreviations of various PDEs are as in the legend of figure 5. b-cam and b-cGS are partial sequences and therefore not numbered.

RNA blot analysis of PDE alpha subunit mRNA.

We have analyzed the complexity of mRNA in the adult bovine, human and mouse retinas. Using a 1.4 kb human cDNA probe (HPA-3, figure 2) to hybridize to human poly (A)+ RNA, a broad band sized from 5.3 to 4.9 kb was identified (figure 7). An apparent single RNA species of 4.5 kb was observed

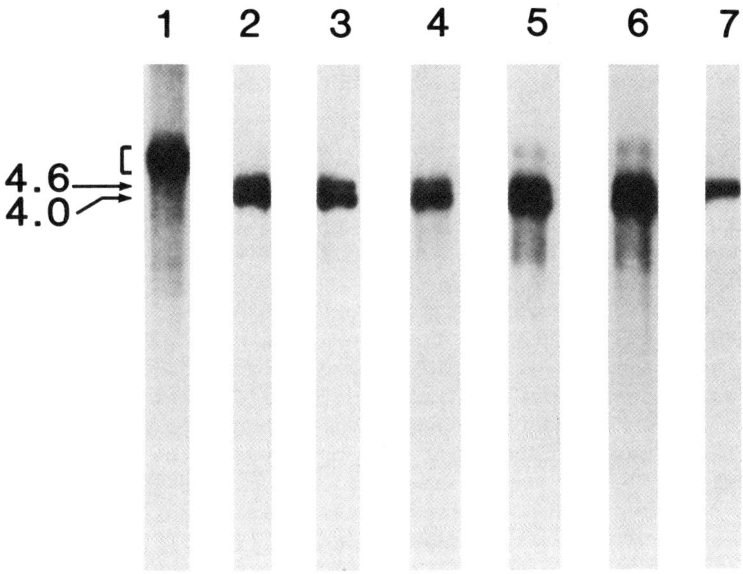

Figure 7. Analysis of human and bovine retinal RNA. Approximately 2 ug retinal poly (A)+ RNA were fractionated on a 1.0% agarose, 2.2 M formaldehyde containing gel and transferred to nitrocellulose (lane 1, human RNA; lanes 2-6, bovine RNA, lane 7, mouse RNA). The results of hybridization with radiolabelled human cDNA clone HPA-3 (figure 2) is shown in lanes 1 and 7. A broad hybridizing band in lane 1 covering a size range from 5.3 to 4.9 kb is bracketed. Hybridization results with bovine cDNA probe 2 (figure 2), are shown in lane 2. Lanes 3-6 show the results of hybridization with oligonucleotide probes A-D (figure 2), respectively.

in mouse retinal RNA with the human cDNA probe (figure 6, lane 7). Several oligonucleotide probes representative of N-terminal and C-terminal region of the alpha-subunit sequence, and a bovine cDNA probe (probe 2, figure 2) hybridize predominantly to two relatively abundant bovine retinal RNAs of 4.6 and 4.0 kb (figure 6, lanes 2-6). Since the length of the open reading frame is identical in the three species, the larger size of the human transcripts must reflect larger 5′ and/or 3′ untranslated regions. The difference in the length of 3′ untranslated regions shown for the human and bovine RNAs (figure 2) suggests that alternative use of poly (A) attachment sites may account for at least some of the observed size heterogeneity.

Gene Structure of the bovine PDE alpha-subunit

We initiated an analysis of the bovine alpha subunit gene to assess the number and size of the gene(s) responsible for transcription of the two bovine RNAs. From the results of genome reconstruction experiments (Pittler et al., 1990) and the ability to identify single genomic restriction fragments with cDNA probes (figure 8, lanes 1, 4, 5) it was determined that there is a single bovine alpha-subunit gene. Since there appears to be a single locus, the size of the gene can be estimated by summing the sizes of nonidentical length BamHI fragments identified with cDNA probes representative of the entire mRNAs (figure 8). We estimated that the bovine gene spans at least 140 kb, which is a minimum value since any restriction fragments that contain only intron sequence would not be detected, and the cDNA clones do not represent the entirety of the alpha subunit mRNAs. However, the possibility that some of the probes hybridized to fragments of closely related genes cannot be formally ruled out.

A preliminary screen of a bovine genomic library yielded a 13.5 kb genomic fragment (G12), of which a 9.5 kb BamHI/SalI fragment strongly hybridized to bovine cDNA. The positions and exon/intron junctions of the seven exons contained in the clone (X1 to X7, figure 9) were determined by DNA sequencing. The exons are short (67bp to 147bp), and separated by six introns of up to 2 kb in length such that 13.5 kb of genomic DNA contain only 750 bp coding sequence. All exon sequences are contained in the domain which is conserved in a wide variety of PDEs from yeast to man

(Pittler et al., 1990; see figure 6) and suspected to harbour the active site of the enzyme. The functional significance of this exon/intron arrangement, if any, is unknown.

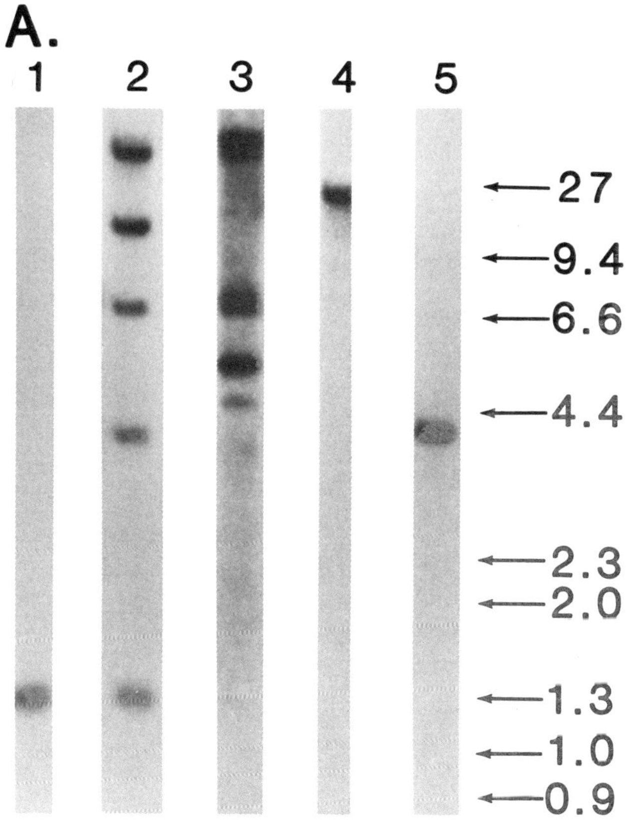

Figure 8. DNA blot analysis of the bovine alpha subunit gene. BamHI genomic DNA fragments identified with cDNA probes. Lanes are designated according to the probe used (see figure 2). Lines between lanes denote bands that represent identical size fragments due to overlap of probes. Size markers are a mixture of lambda DNA digested with HindIII and HaeIII digested phiX174 DNA (BRL).

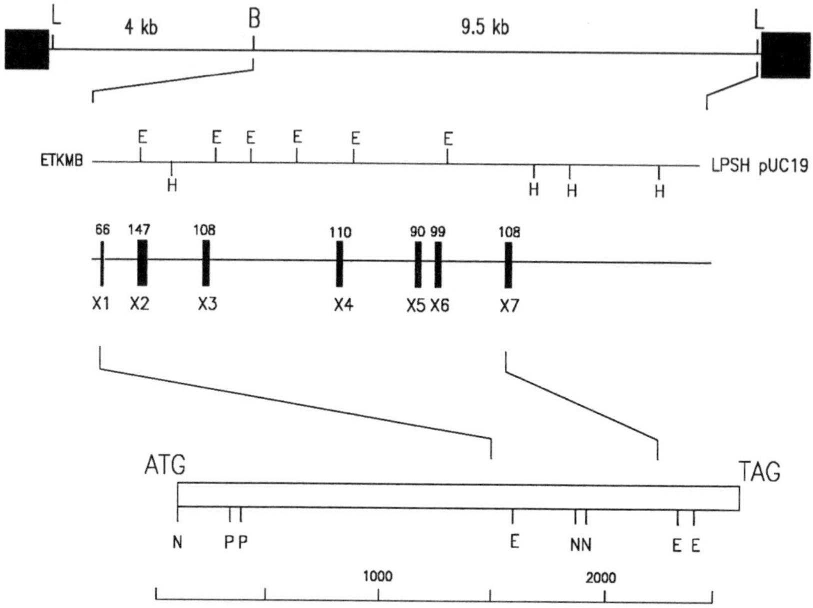

Figure 9. Exon distribution in a bovine PDE alpha-subunit genomic clone (BPA-G12). Black rectangles depict exons (X1 - X7). The length of each exon is indicated in bp above the rectangles. The box delimited by ATG and TAG symbolizes the the alpha-subunit coding sequence. The 4kb SalI-BamHI fragment of the genomic clone did not hybridize to alpha-subunit cDNA fragments (intron sequences only). E, EcoRI; L, SalI; B, BamHI; H, HindIII, P, PstI; N, NcoI. ETKMB, LPSH: multiple cloning sites of pUC19.

The potential for isozymes to be encoded by the gene, its large size, and complex exon-intron organization make it comparable to the Drosophila dunce gene, another member of the PDE gene family. The dunce gene spans more than 100kb of DNA, encodes at least six different RNAs, and contains idependent transcription units within a relatively large intron in the upstream region (Davis and Davidson, 1986; Chen et al. 1986, 1987). It is intriguing to speculate that the apparent complexity of these two genes may be a general feature of PDE genes within this gene family.

The rod PDE beta-subunit

The beta-subunit is biochemically and biophysically closely related to the alpha-subunit. Their mobilities in denaturing polyacrylamide gels (SDS PAGE) are identical under some conditions, and slightly different under specific crosslinking conditions (Baehr et al., 1982). Limited proteolysis of labeled, gel purified alpha-and beta-subunits yields similar peptide patterns. Preparative methods for purification of either alpha or beta under denaturing conditions, apart from gel electrophoresis in the presence of SDS, have failed. A partial amino acid sequence for the beta-subunit, based on cDNA sequencing (75 residues, homologous to pos. 536 to 610 of the alpha-subunit, figure 4) has been published (Ovchinnikov et al., 1987). The partial sequence is 73% identical to the corresponding part of the alpha-sequence and covers part of a conserved domain common to a number of cyclic nucleotide phosphodiesterases (figure 5).

The rod PDE gamma-subunit

Bovine rod PDE was shown to copurify with a small gamma-subunit of approximately 13K mobility (Baehr et al., 1979). Subsequently, the gamma-subunit was purified to homogeneity by taking advantage of its heat and acid stability, and shown to act as an internal inhibitor that is sensitive to trypsin (Hurley and Stryer, 1982). The complete amino acid sequence of the bovine gamma-subunit was determined by both peptide and cDNA sequencing (Ovchinnikov et al., 1986), and its molecular weight calculated as 9,700. The N-terminus appears to be blocked, and no other known posttranslational processing has been identified. The gamma-subunit is an 87 residue hydrophilic polypeptide with a basic core domain (10 positively charged residues at pos. 24 - 45, figure 4), presumably the site of trypsin sensitivity (Lipkin et al., 1988). Competition assays with synthetic peptides (Sakamoto et al., 1989) indicate that this basic core may also interact with the alpha-subunit of transducin.

Recently the mouse (Tuteja and Farber, 1988; Danciger et al., 1989) and human (Ringens et al., 1989; Farber et al., 1989) PDE gamma-subunit were cloned and sequenced. Except for two residues near the N-terminus, the mouse, human and bovine gamma-subunits are identical (figure 10).

The only amino acid being different in all three species, is that at position 17.

```
                ↓
MNLEPPKAEIRSATRVMGGPVTPRKGPPKFKQRQTRQFKSKPPKKGVQG  b
       G        I                                  m
         F      G                                  h

FGDDIPGMEGLGTDITVICPWEAFNHLELHELAQYGII  b
                                        m
                                        h
```

Figure 10. Amino acid sequences of the rod PDE gamma-subunit of bovine (b), mouse (m), and human (h). Amino acid position 17 is marked with an arrow, the basic core domain (pos. 24 - 45) is underlined.

Two recombinant constructs encoding gamma subunit fusion proteins which are expressed at high levels in bacterial systems have been reported. In one construct (Brown and Stryer, 1989) a synthetic bovine gamma subunit cDNA sequence was cloned into an expression vector that uses the lambda P_L promoter. The expressed fusion protein can be cleaved with serum clotting factor Xa to yield a peptide that is indistinguishable from the native gamma subunit. Both the fusion peptide and the gamma subunit generated by proteolysis of the fusion peptide inhibit PDE and interact with transducin. In another construct (Griswold-Prenner et al., 1989) mouse gamma cDNA was cloned into the widely used pUC19 plasmid vector (lacZ promoter). The expressed fusion protein consisting of 18 additional residues encoded by the pUC19 polylinker sequence also inhibited trypsin activated PDE. Expressed gamma lacking the C-terminal domain (13 residues) did not inhibit, but instead weakly stimulated, cGMP hydrolysis (Brown and Stryer, 1989), consistent with an earlier finding (Sakamoto et al., 1989) that the C-terminal domain of gamma contains the PDE binding site.

Chromosomal localization of PDE genes

The continuing development of the human gene map and synteny relationships among mammals is essential to expedite our ability to identify and study defects in genes that are primary causes of hereditary disorders. Extending this effort toward the localization of retina specific genes

(Blatt et al., 1988, Nathans, 1987, Liou et al., 1989), the alpha-subunit gene was mapped to the region 5q31.2-q34 (Pittler et al., 1990), and the gamma subunit gene to the long arm of chromosome 17 (Pittler, S. J. and Ledbetter, D., unpublished results). In mouse, the alpha subunit gene was shown to be on chromosome 18 (Deborah Farber, personal communication), and the gamma subunit gene on chromosome 11 (Danciger et al., 1989).

Clinical implications of a defective PDE

Approximately twenty-five percent of known inherited metabolic defects affect the eye, yet in virtually all of these disorders the cause is unknown. Many of the known genetic eye disorders manifest specifically in the retina. Since these defects are generally nonlethal it is not very often possible to obtain and study the retinas at an early stage of the disease. Therefore, research efforts have been intensely focused on the study of retinal degeneration in animals. A great deal of evidence has been reported that suggests a link between retinal degeneration and defects in phototransduction metabolism. Hollyfield and coworkers (Lolley et al., 1976; Ulshafer et al.; 1980; Hollyfield et al., 1982, 1985) demonstrated that elevated levels of cGMP are toxic to photoreceptor cells in Xenopus laevis eye rudiments, and in the adult human retina maintained in culture. There is a marked specificity of the degeneration for rod cells. Several animal retinal degeneration mutants have been identified which exhibit very high levels of photoreceptor cGMP (Farber and Lolley, 1974; Aguirre et al., 1978; Woodford et al., 1982; Leon et al., 1988), prior to the onset of any morphological disruption. In one of these mutants, the rd mouse (Brückner, 1951), cGMP PDE has been shown to exhibit abnormal subunit assembly (Lee et al., 1988), and to display altered kinetic properties (Lolley and Lee, 1984). The photoreceptor layer of these mice undergo complete degeneration within 20 days postnatally (Noell, 1958). Recently isolated candidate rd cDNAs and the absence of linkage of the alpha and gamma genes, however, did not suggest involvement of cGMP PDE in the primary cause of the disorder (Bowes and Farber, 1989; Pittler and Ledbetter, unpublished results).

An inherited rod-cone dysplasia in the Irish Setter dog shows very similar characteristics to that observed in the rd mouse (Aguirre et al., 1978). Within 18 to 20 weeks postnatally there is a complete loss of rod photoreceptor cells with only a sparse number of cones remaining. Levels of cGMP rise as high as 10 fold above normal prior to the onset of morphological degeneration. However, results from biochemical analysis suggest that the cause of the disorders differ substantially (Lee et al., 1985; Lolley et al., 1987) The PDE complex in the rd mouse retina does not form properly while that of Irish Setters appears normal in subunit assembly. PDE activity appears inactive *in vivo* in contrast to the enzyme in the rd retina. It is possible that a point mutation in one of the PDE subunits which does not alter the catalytic activity but abolishes the ability to interact with other components of the phototransduction cascade is the cause of the disorder. Further analysis of PDE subunit assembly and membrane interaction will be required to understand the molecular basis of these animal disorders.

Recently, high cGMP levels were reported in rod photoreceptors of a 17 year old human male (Farber et al., 1987) with autosomal dominant retinitis pigmentosa (ADRP). The authors suggested a strong correlation in morphological and biochemical characteristics in this patient's retina, compared to that observed in the rd mouse. Indeed, the cGMP PDE gave rise to abnormal kinetic parameters similar to the rd mouse enzyme. Moreover, the morphological degeneration observed was shown to be mostly limited to areas of the photoreceptor layer that are predominantly composed of rod cells. ADRP is one of a group of clinically related retinal dystrophies for which there is presently no treatment and the cause is unknown (For review, see Heckenlively, 1988). Recently, a four year study of a large Scottish family carrying the ADRP gene yielded strong linkage with RFLPs identified on chromosome 3 (Peter Humphries and Stephen Daiger, personal communication) in the vicinity of the gene encoding rhodopsin.

Since the metabolism of cGMP in photoreceptors involves complex interactions of several proteins, the cause of the retinal degenerations described may be due to specific defects in any of the proteins of the phototransduction cascade. Biochemical analysis of rhodopsin, transducin, and arrestin in rd mice have shown that these proteins operate normally in the mutant retina. Linkage analyses using rhodopsin, transducin, arrestin, IRBP, and CRALB cDNAs have not detected linkage in the RP families studied so far. The potential involvement of a defect in rod PDE in the etiology of human retinal dystrophies is currently being investigated by several laboratories.

SUMMARY

Metabolism of cGMP is critically important for the functioning of phototransduction in the mammalian retina. In rod and cone photoreceptors, two types of antagonistic enzymes, guanylate cyclases and cGMP phosphodiesterases, carefully balance the available amount of the intracellular messenger. Guanylate cyclase produces cGMP and phosphodiesterase rapidly hydrolyzes cGMP upon bleaching of the photopigment. Regulation of their activity in light and dark, influence of Ca^{++}, and feed-back mechanisms are currently under intense investigation. A molecular analysis on both the gene and protein levels will contribute significantly to our understanding of their respective roles in phototransduction.

The two types of enzymes have been characterized molecularly to a very different extent. Rod phosphodiesterase was purified to homogeneity almost fifteen years ago, but photoreceptor guanylate cyclase has evaded all attempts for molecular characterization. Characterization of retinal guanylate cyclase cDNA(s), however, will most likely be achieved in the near future. Cone PDE was shown to be a distinct enzyme, different from, but related to, the rod enzyme. Molecular cloning has provided sequence information of two of the three subunits of rod PDE; the small inhibitory subunit has been expressed in bacterial expression vectors, giving us an elegant tool for exploring mechanisms of activation and inhibition. The gene encoding the alpha subunit was shown to be a member of a large gene family of cyclic nucleotide phosphodiesterases, present in many eucaryotes ranging from unicellular organisms (yeast) to mammals.

While much has been achieved, many questions remain to be answered. The beta subunit of rod phosphodiesterase has evaded complete molecular characterization, and its origin (one gene and posttranslational modification of the gene product generating alpha and beta, alternative splicing, or two separate genes with distinct gene products) has not been elucidated. Mechanisms of interaction of subunits, activation and inhibition, the active site(s) of the enzyme are undefined. Virtually nothing is known about the molecular organization of the photoreceptor guanylate cyclase(s). Recent cloning of two apparently unrelated mammalian guanylate cyclases, however, containing a common homologous domain signals increasingly rapid progress in this field.

REFERENCES

Aguirre G et al. (1978). Rod-cone dysplasia in Irish Setters: a defect in cyclic GMP metabolism in visual cells. Science 201:1133-1134.

Ashman et al. (1963). Isolation of adenosine 3′,5′-monophosphate and guanosine 3′,5′-monophosphate from rat urine. Biochem Biophys Res Commun 11:330-334.

Baehr W, Devlin M, Applebury ML (1979). Isolation and characterization of cGMP phosphodiesterase from bovine rod outer segemnts. J Biol Chem 254:11669-11677.

Baehr W et al. (1982). Characterizaction of bovine rod outer segment G-protein. J Biol Chem 252:6452-6460.

Baehr W, Pittler SJ (1989). Rod photoreceptor cGMP phosphodiesterase alpha-subunit of the mouse retina: cDNA sequence and intergenic comparison to human and bovine. Exp. Eye Research (submitted).

Barbehenn EK et al. (1985) Extracellular cGMP phosphodiesterase related to the rod outer segment phosphodiesterase isolated from bovine and monkey retinas. Biochemistry 24:1309-1316.

Beavo JA (1988) Multiple Isozymes of cyclic Nucleotide Phosphodiesterase. Adv Second Messenger and Phosphoprotein Research 22:1-38.

Blatt C et al. (1988). Chromosomal localization of genes encoding guanine nucleotide-binding protein subunits in mouse and human. Proc Natl Acad Sci USA 85:7642-7646.

Bowes C, Farber DB (1989). Isolation and partial characterization of mouse photoreceptor specific cDNAs - a strategy for the isolation of candidate rd cDNAs. Inv. Ophth. Vis Sci. Suppl. 30:113.

Brown RL, Stryer L (1989). Expression in bacteria of functional inhibitory subunit of retinal rod cGMP phosphodiesterase. Proc Natl Acad Sci USA 86:4922-4926.

Brückner R (1951). Spaltlampenmikroskopie und Ophthalmoskopie am Auge von Ratte und Maus. Document. Ophthalm. 5-6:452-554.

Chabre M, Deterre P (1989). Molecular mechanism of visual

transduction. Eur J Biochem 179:255-266.
Charbonneau H et al. (1986). Identification of a conserved domain among cyclic nucleotide phosphodiesterases from diverse species. Proc Natl Acad Sci USA 83:9308-9312.
Chen C-N et al. (1986). Molecular analysis of cDNA clones and the corresponding genomic coding sequences of the Drosophila dunce$^+$ gene, the structural gene for cAMP phosphodiesterase. Proc Natl Acad Sci USA 83:9313-9317.
Chen C-N et al. (1987). At least two genes reside within a large intron of the *dunce* gene of Drosophila. Nature 329:721-724.
Chinkers M et al. (1989). A membrane form of guanylate cyclase is an atrial natriuretic peptide receptor. Nature 338:78-83.
Clarke S et al. (1989). Posttranslational modification of the Ha-ras oncogene protein: evidence for a third class of protein carboxyl methyltransferases. Proc Natl Acad Sci USA 85:4643-4647.
Colicelli J et al. (1989). Isolation and characterization of a mammalian gene encoding a high affinity cAMP phosphodiesterase. Proc Natl Acad Sci USA 86:3599-3603.
Danciger M et al. (1989). The gene for the gamma-subunit of retinal cGMP phosphodiesterase is on mouse chromosome 11. Exp Eye Res 48:303-308.
Davis RL, Davidson N (1986). The memory gene *dunce*$^+$ encodes a remarkable set of RNAs with internal heterogeneity. Mol Cell Biol 6:1464-1470.
Davis RL et al. (1989). Cloning and characterization of mammalian homologs of the Drosophila dunce gene. Proc Natl Acad Sci USA 86:3604-3608.
Deterre P et al. (1988). cGMP phosphodiesterase of retinal rods is regulated by two inhibitory subunits. Proc Natl Acad Sci USA 85:2424-2428.
Falk JD, Applebury ML (1988). The molecular genetics of photoreceptor cells. Progress in Retina Research 7:89-112.
Farber DB, Lolley RN (1974). Cyclic guanosine monophosphate: elevation in degenerating photoreceptor cells of the C3H mouse retina. Science 186:449-451.
Farber DB et al. (1987). Histopathological and biochemical studies on donor eyes affected with retinitis pigmentosa. In Hollyfield JG, Anderson RA, LaVail MM (eds): "Degenerative Retinal Disorders: Clinical and Laboratory Investigations", New York, Alan R. Liss, pp. 53-67.
Farber DB et al. (1989). Cloning and sequencing of the human retinal cGMP phosphodiesterase gamma-subunit. Inv

Opth Vis Sci Suppl 30:115.
Fesenko EE, Kolesnikov SS, and Lyubarsky AL (1985). Induction by cyclic GMP of cationic conductance in plasma membrane of retinal rod outer segment. Nature 313:310-313.
Fleischman D, Denisevich M (1979). Guanylate Cyclase of Isolated Bovine Retinal Rod Axonemes. Biochemistry 18: 5060-5066.
Fung BK-K, Stryer L (1980). Photolyzed rhodopsin catalyzes the exchange of GTP for bound GDP in retinal rod outer segments. Proc Natl Acad Sci USA 77:2500-2504.
Fung BK-K, Hurley JG, Stryer L 1981. Flow of information in the light-triggered cyclic nucleotide cascade of vision. Proc Natl Acad Sci USA 78:152-156.
Fung BK-K, Griswold-Prenner I (1989). G protein-Effector coupling: Binding of rod phosphodiesterase inhibitory subunit to transducin. Biochemistry 28:3133-3137.
Gillespie PG, Beavo JA (1988). Characterization of a bovine cone photoreceptor phosphodiesterase purified by cyclic GMP-Sepharose chromatography. J Biol Chem 263:8133-8141.
Gillespie PG, Beavo JA (1989). cGMP is tightly bound to bovine retinal rod phosphodiesterase. Proc Natl Acad Sci USA 86:4311-4315.
Goldberg ND, O'Dea RF, Haddox MK (1973). cyclic GMP. Adv Cycl Nucl Res 3:155-223.
Goldberg ND, Haddox MK (1977). Cyclic GMP metabolism and involvement in biological regulation. Ann Rev Biochem 46:823-896.
Griswold-Prenner I et al. (1989). G protein-effector coupling: Interactions of recombinant inhibitory gamma subunit with transducin and phosphodiesterase. Biochemistry 28:6145-6150.
Gutierrrez L et al. (1989). Post-translational processing of $p21^{ras}$ is two-step and involves carboxyl-methylation and carboxy-terminal proteolysis. EMBO J 8:1093-1098.
Haynes L, Yau K-W (1985). Cyclic GMP-sensitive conductance in outer segment membrane of catfish cones. Nature 317:61-64.
Hancock J et al. (1989). All *ras* proteins are polyisoprenylated but only some are palmitoylated. Cell 57:1167-1177.
Heckenlively JR (1988). In "Retinitis Pigmentosa" JB Lippincott Co., Philadelphia.
Hollyfield JG et al. (1982). Selective photoreceptor degeneration during retinal development: the role of altered phosphodiesterase activity and increased levels of cyclic nucleotides. In Hollyfield JG (ed):

"Structure of the eye", Elsevier-North Holland, p.97.
Hollyfield JG et al. (1985). Simulation of an inherited disease in the normal retina: involvement of cyclic nucleotides. In LaVail MM, Hollyfield JG, Anderson RE (eds): "Retinal Degeneration: Experimental and Clinical Studies", Alan R. Liss, New York, pp. 341-348.
Hughes SM, Brand MD (1983). Size changes of phosphodiesterase in bovine rod outer segments on illumination. Biochemistry 22:1704-1708.
Hurley JB (1987). Molecular properties of the cGMP cascade of vertebrate photoreceptors. Ann Rev Physiol 49:793-812.
Hurley JB, Stryer L (1982). Purification and characterization of the gamma-regulatory subunit of the cyclic GMP phosphodiesterase from retinal rod outer segments. J Biol Chem 257:11094-11099.
Hurwitz RL, Bunt-Milam AH, Beavo JA (1984). Immunologic characterization of the photoreceptor outer segment cGMP phosphodiesterase. J Biol Chem 259:8612-8618.
Hurwitz RL et al. (1985). cGMP phosphodiesterase in rod and cone outer segments of the retina. J Biol Chem 260:568-573.
Koch KW, Stryer L (1988). Highly cooperative feedback control of retinal rod guanylate cyclase by calcium ions. Nature 334:64-66.
Koesling D et al. (1988). The primary structure of the 70 kDa subunit of bovine soluble guanylate cyclase. FEBS Letters 239:29-34.
Kühn H (1980). Light- and GTP-regulated interaction of GTPase and other proteins with bovine photoreceptor membranes. Nature 283:587-589.
Lee RH et al. (1985). Phosphodiesterase probes show distinct defects in rd mice and Irish setter dog disorders. Invest Opthalmol Vis Sci 26:1569-1579.
Lee RH et al. (1988). Characterization of a phosphodiesterase-immunoreactive polypeptide from rod photoreceptors of developing rd mouse retinas. Invest. Ophthalmol. Visual Sci. 29:1021-1027.
Leon A, Hussain AA, Curtis R (1988). Rod-cone dysplasia in the Rdy cat: electrophysiology and biochemistry. Opthalmol Vis Sci Suppl 29:143.
Lerea CL et al. (1986). Identification of specific transducin alpha-subunits in retinal rod and cone photoreceptors. Science 234:77-80.
Liebman PA, Parker KR, Dratz EA (1987). The molecular mechanism of visual excitation and its relation to the structure and composition of the rod outer segment. Ann

Rev Physiol 49:765-791.
Liou GL et al. (1989). Human interphotoreceptor retinoid-binding protein: gene structure and primary sequence. J Biol Chem 264:8200-8206.
Lipkin VM et al. (1988) Active sites of the cGMP phosphodiesterase gamma-subunit of retinal rod outer segments. FEBS Letters 234:287-290.
Lolley RN et al. (1976). Cyclic GMP accumulation causes degeneration of photoreceptor cells: simulation of an inherited disease. Science 196:664-666.
Lolley RN, Lee RH (1984). Phosphodiesterase dysfunction, cyclic GMP accumulation, and visual cell degeneration in early-onset inherited blindness. Adv Cycl Nucl Prot Phos Res 17:315-327.
Lolley RN et al. (1987). Inherited disorders of rd mice and affected irish setter dogs: Evaluation of transducin and cGMP phosphodiesterase. In "Degenerative Retinal Disorders: Clinical and Laboratory Investigations", Alan R. Liss, New York, pp. 269-287.
Magee T, Hanley M (1988). Sticky Fingers and CAAX boxes. Nature 335:114-115.
Miki N et al. (1975). Purification and properties of the Light-activated cyclic nucleotide phosphodiesterase of rod outer segments. J Biol Chem 250:6320-6327.
Nathans J, Hogness DS (1983). Isolation, sequence analysis, and intron-exon arrangement of the gene encoding bovine rhodopsin. Cell 34:807-814.
Nathans J, Thomas D, Hogness DS (1986). Molecular genetics of human colour vision: The genes encoding blue, green, and red pigments. Science 232:193-202.
Nathans J (1987). Molecular Biology of Visual Pigments. Ann Rev Neuroscience 10:163-194.
Newmark P (1989). Guanylate cyclases send out new signals. Nature 338:18.
Noell WK (1958). Differentiation, metabolic organization, and viability of the visual cell. AMA Arch Ophthalmol 60:702-733.
Ovchinnikov YA et al. (1986). cyclic GMP phosphodiesterase from cattle retina: Amino acid sequence of the gamma-subunit and nucleotide sequence of the corresponding cDNA. FEBS Letters 204:288-292.
Ovchinnikov YA et al. (1987). Cyclic cGMP phosphodiesterase from bovine retina: Amino acid sequence of the alpha-subunit and nucleotide sequence of the corresponding cDNA. FEBS Letters 223:169-173.
Pannbacker RG (1973). Control of guanylate cyclase activity

in the rod outer segment. Science 182:1138-1140.

Pannbacker RG, Lovett K (1977). Localization of cyclic nucleotide phosphodiesterase activity within the bovine photoreceptor cell. Invest Ophthtalmol Vis Sci 16:166-168.

Pittler SJ, McConnell DG, Davis RL (1987). Am J Hum Genet Suppl A234.

Pittler SJ et al. (1988). J Cell Biology 107:A3638 (replacement).

Pittler SJ et al. (1990). Molecular characterization of human and bovine rod photoreceptor cGMP phosphodiesterase alpha-subunit and chromosomal localization of the human gene. Genomics, in press.

Pugh EN, Cobbs WH (1986). Visual transduction in vertebrate rods and cones: a tale of two transmitters, calcium and cGMP. Vision Res 26:1613-1643.

Ringens PJ et al. (1989). Analysis of DNA from Retinitis pigmentosa patients with cloned candidate retinal genes. Inv Opth Vis Sci Suppl 30:43.

Sakamoto K et al. (1989). Identification of the multifunctional sites of the phosphodiesterase gamma subunit. Inv Ophth Vis Sci Suppl 30:172.

Sass et al. (1986). Cloning and characterization of the high affinity cAMP phosphodiesterase of _Saccharomyces cerevisiae_. Proc Natl Acad Sci USA 83:9303-9307.

Schulz S, Chinkers M, Garbers DL (1989). The guanylate cyclase/receptor family of proteins. FASEB J 3:2026-2035.

Stryer L (1985). Molecular design of an amplification cascade in vision. Biopolymers 24:29-47.

Stryer L (1986). Cyclic GMP cascade of vision. Ann Rev Neurosci 9:87-119.

Swanson RJ, Applebury ML (1983). Methylation of Proteins in Photoreceptor Rod Outer segments. J Biol Chem 258:10599-10605.

Takemoto LJ et al. (1984). cyclic GMP phosphodiesterase from bovine retina: Evidence for interspecies conservation. Biochem J 217:129-133.

Takemoto D et al. (1982) Peptide mapping of multiple forms of cyclic nucleotide phosphodiesterase. J Biol Chem 257:14597-14599.

Tremblay J, Gerzer R, Hamet P (1988). Cyclic GMP in cell function. Adv in Second Messenger and Phosphoprotein Res 22:319-383.

Tuteja N, Farber DB (1988). Gamma-subunit of mouse retinal cGMP phosphodiesterase: cDNA and corresponding amino acid sequence. FEBS Letters 232:182-186.

Uchida S et al. (1981). A GTP-protein activator of

phosphodiesterase which forms in response to bleached rhodopsin. J Cyclic Nuc Res 7:95-104.
Ulshafer R, Garcia C, Hollyfield JG (1980). Sensitivity of photoreceptors to elevated levels of cGMP in the human retina. Invest Ophthalmol Vis Sci 19:1236-1241.
Wensel TG, Stryer L (1986). Reciprocal control of retinal rod cyclic GMP phosphodiesterase by its gamma-subunit and transducin. PROTEINS: Structure, Function, and Genetics 1:90-99.
Whalen MM, Bitensky MW (1989). Comparison of the phosphodiesterase inhibitory subunit interactions of frog and bovine outer segments. Biochem J 259:13-19.
Woodford B et al. (1982). Cyclic nucleotide metabolism in inherited retinopathy in collies: a biochemical and histochemical study. Exp Eye Res 34:703-714.
Yamazaki A et al. (1983). Activation mechanism of rod outer segment cyclic GMP phosphodiesterase. Release of inhibitor by the GTP/GDP-binding protein. J Biol Chem 258:8188-8194.
Yee R, Liebman PA (1978). Light-activated phosphodiesterase of the rod outer segments. Kinetics and parameters of activation and deactivation. J Biol Chem 253:8902-8909.

Molecular Biology of the Retina: Basic and Clinically Relevant Studies, pages 67–86

STUDIES LEADING TO THE ISOLATION OF A cDNA FOR THE GENE CAUSING RETINAL DEGENERATION IN THE *rd* MOUSE

Debora B. Farber, Cathy Bowes
and Michael Danciger*

Jules Stein Eye Institute and Department of Ophthalmology, UCLA School of Medicine, and *Loyola Marymount University, Los Angeles, California 90024-7008

The primary lesion affecting most inherited retinal diseases in animals and humans is unknown, but biochemical defects in cyclic nucleotide metabolism have been associated with some of these disorders (Farber and Lolley, 1974; Farber and Lolley, 1977; Aguirre et al., 1978; Woodford et al., 1982; Farber et al. 1987). In this chapter we will first discuss the involvement of elevated levels of cyclic GMP in the genetic retinal degeneration of the *rd* mouse, the animal model that we have most thoroughly studied. Secondly, we will describe the work that we have carried out for the identification and isolation of a candidate cDNA for the *rd* gene.

The *rd* disease has a very early onset and a fast progression. LaVail and Sidman (1974) showed with light microscopy that at 10 postnatal days there is no morphological difference between the normal and *rd* photoreceptors. However, by 13 postnatal days the *rd* visual cells are degenerating rapidly and by 20 days of age, most of the photoreceptor cells have died. Only one row of pyknotic nuclei corresponding to cone visual cells remains in the outer nuclear layer for a few more postnatal days (Carter-Dawson et al., 1978). By 30 days, the retinal pigment epithelium is in direct apposition to the cells of the inner retina which survive the disease. At the electron microscopic level, the first pathological changes are observed at 8 days of age. These include disorganization of the disk membranes of the outer segments and swelling of the mitochondria of the inner segments (Lasansky and DeRobertis, 1960; Sonohara and Shiose, 1968).

Many years ago, we found that elevated levels of cyclic GMP are present in the *rd* retina by postnatal day 6, before any pathology is observed, and that the accumulation of cyclic GMP is restricted to the photoreceptor cells (Farber and Lolley, 1974). High levels of cyclic GMP could result either from an increase in its synthesis or from a decrease in its degradation. We found that the activity of guanylate cyclase, the enzyme which synthesizes cyclic GMP, is normal in *rd* visual cells while they are viable (Farber and Lolley, 1976), whereas the activity of cyclic GMP-phosphodiesterase, the enzyme which hydrolyzes cyclic GMP, is much lower than normal from a very early age (Farber and Lolley, 1976). This deficient activity is restricted to the photoreceptors of the *rd* retina. Other body tissues of the *rd* mouse including blood (plasma, serum, lymphocytes and erythrocytes), muscle, brain and skin have cyclic GMP-phosphodiesterase activities comparable to those of control mouse tissues (Lolley and Farber, 1980).

Cyclic GMP-phosphodiesterase is a key enzyme in the phototransduction process of normal rod photoreceptors. The enzyme is composed of four subunits, two large ones with catalytic activity, α and β (88 and 84 kDa, respectively) and two small γ subunits (10 kDa) with inhibitory characteristics (Baehr et al., 1979). Light activation of cyclic GMP-phosphodiesterase is a complex process which requires the action of some components of the rod outer segments such as photolyzed, unphosphorylated rhodopsin, GTP and a G-protein called transducin (Fung et al., 1981). Transducin is a trimeric protein, composed of α, β and γ subunits. Similar to the activation by light, the deactivation of cyclic GMP-phosphodiesterase is also complex (Pfister et al., 1983) and requires the participation of other components of the photoreceptors like ATP, rhodopsin kinase (which phosphorylates rhodopsin) and a 48-kDa protein, also known as S-antigen or arrestin.

Many factors could contribute to an abnormal activity of cyclic GMP-phosphodiesterase in the *rd* retina. These may include changes in the structure of the enzyme, the presence of a reduced number of enzyme molecules, the action of inhibitory elements, or a lesion in any of the proteins involved in the cyclic GMP-phosphodiesterase activation/deactivation. We have investigated each of these factors in normal and *rd* retinas.

In order to determine whether the enzyme is produced normally by the photoreceptor cells, we studied its biosynthesis in *rd* retina and found that it is comparable to that in normal retina: synthesis starts in both normal and *rd* tissues at 6 postnatal days and the same amount of enzyme is made in both cases (Farber et al., 1988). Thus, the production of the enzyme is not defective prior to the degeneration of the photoreceptor cells. However, the levels of cyclic GMP-phosphodiesterase present in the *rd* retina are lower than those in control retina from very early in development, suggesting that the enzyme is more labile and gets degraded very soon after it is made. This may be due to a change in the composition of the enzyme itself or to the presence of a more active proteolytic system in the *rd* retina compared to control retinas (Farber et al., 1988). Preliminary studies seem to indicate that the latter is not the case.

Although cyclic GMP-phosphodiesterase is present in the *rd* retina, only a small fraction of the retinal *rd* cyclic GMP is hydrolyzed in response to light while the major portion of cyclic GMP remains intact (Farber and Lolley, 1977). This indicates that the *rd* enzyme is minimally activated by light and, therefore, strengthens the possibility of a lesion in the cyclic-GMP-phosphodiesterase itself or in any of the components of the rod outer segment that are involved in its activation / deactivation.

In our investigations on the functionality of the rhodopsin, from *rd* retina, we found that there is an abnormality in its phosphorylation/dephosphorylation. The concentration of rhodopsin is comparable in normal and *rd* retinas up to 11- 12 days at which time it starts to decrease in the *rd* retina due to the loss of the photoreceptors. However, rhodopsin can not incorporate ^{32}P from ATP *in vitro* at any age in the *rd* retina (Shuster and Farber, 1986). The possible culprits for this abnormality are either rhodopsin kinase, which catalyzes the transfer of phosphate from ATP to rhodopsin, rhodopsin itself, or the phosphatase that dephosphorylates rhodopsin, protein phosphatase 2A. Preliminary studies done in collaboration with Krzysztof Palczewski and Paul Hargrave (1990) indicate that rhodopsin kinase and rhodopsin are normal in *rd* retina whereas the *rd* protein phosphatase 2A has a much higher activity than the corresponding enzyme from control retina. This could explain the lack of *in vitro* phosphorylation of *rd* rhodopsin, since the elevated level of protein phosphatase 2A activity could remove the phosphate from rhodopsin as soon as the visual pigment is phosphorylated. Previous investigations examining the effect of several cations and nucleotides on the phosphorylation of rhodopsin in rod outer segments isolated from normal retinas, showed that high concentrations of cyclic GMP inhibit the incorporation of ^{32}P into rhodopsin (Shuster and Farber, 1984). These experiments suggested that the inhibition of rhodopsin phosphorylation could result from an enhancement of the phosphoprotein phosphatase activity by cyclic GMP. It is highly probable that this occurs in the *rd* retina, a hypothesis which we are currently testing. This abnormality in the *rd* retina would then be secondary to the cyclic GMP-phosphodiesterase deficiency that generates the high cGMP levels.

Another approach that we used to look for abnormalities in the proteins involved in cyclic GMP-phosphodiesterase activation/deactivation processes in *rd* retina was to analyze the mRNAs that code for those proteins in control and *rd* mice with regard to presence, size, concentration and time of onset of expression. The following probes were used: opsin cDNA, obtained from Jeremy Nathans, (Johns Hopkins), the cDNAs for the α, β and γ subunits of transducin, given to us by Melvin Simon (California Institute of Technology) and 48-kDa protein cDNA, a gift from Toshimichi Shinohara (National Eye Institute).

With these cDNAs we probed Northern blots which contained RNAs from the retinas of 10-11-day-old homozygous *rd* and heterozygous *rd*/+ littermates, and adult, wild type, normal mice. RNAs from mouse brain, spleen and liver were

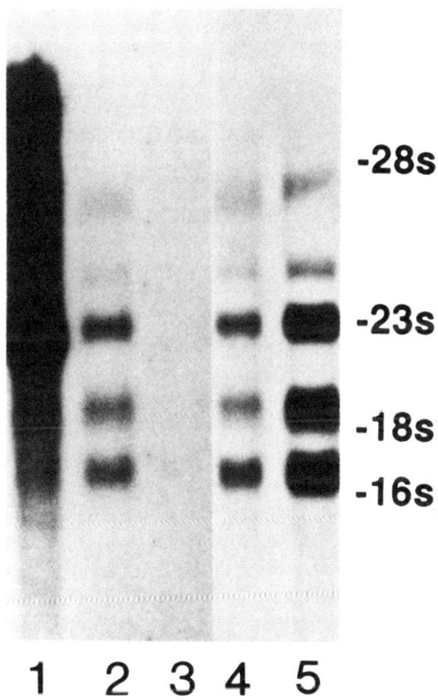

Figure 1. Hybridization of opsin cDNA to mRNAs from retinas and liver. Lane 1: bovine retinal RNA. Lane 2: 10-11-day-old *rd* / *rd* mouse retinal RNA. Lane 3: adult *rd* / + mouse liver RNA. Lane 4: 10-11-day-old *rd* / + mouse retinal RNA. Lane 5: adult +/ + retinal RNA. The sizes of ribosomal RNA markers shown on the right are: 28s, 4.72kb; 23s, 2.9 kb; 18s, 1.87 kb; 16s, 1.54 kb. (From Bowes and Farber, 1987, with permission of the publisher.)

also run as negative controls. In addition, another cDNA probe, pCHO B, which hybridizes a specific mRNA constitutively expressed in all mammalian tissues (Harpold et al., 1979), was used to confirm that comparable amounts of RNA were loaded in all lanes of the Northern blots. Figure 1 is an example of an autoradiogram of a Northern blot hybridized with a cDNA coding for opsin: both control and *rd* retina gave five bands with similar size and signal intensity.

The same kind of results were obtained with all of the probes used: each hybridized comparable mRNAs from normal and *rd* retinas. Thus, the mRNAs coding for opsin, the three subunits of transducin and the 48-kDa protein are present in the *rd* retina, and they have the same apparent size and are expressed in concentrations qualitatively similar to those found in normal retina (Bowes and Farber, 1987).

In addition, we investigated whether the time of onset of expression of these mRNAs, as well as the time of onset of expression of the translated proteins, is the same in the normal and degenerative retinas, because a delay in the appearance and action of any of the mRNAs and/or proteins could be associated with the etiology of the *rd* disease (Bowes et al., 1988). ^{32}P-labeled cDNA probes were hybridized to slot-blotted RNA from *rd*/*rd* and *rd*/+ mice isolated at each postnatal day during development. Figure 2 is an example of the autoradiograms obtained; in this particular case, an opsin cDNA probe was used to hybridize the immobilized RNA. In order to obtain quantitative data, the autoradiograms were then analyzed by laser densitometry.

Figure 3 shows the quantitation of the levels of opsin mRNA as a function of postnatal development. Expression of opsin mRNA is first detectable, although at minimal levels, at postnatal day 5 in both *rd* and control retina; its concentration is comparable in the two types of retina until day 11 and then it begins to decrease in the *rd* retina following the degeneration of the photoreceptor cells. With regards to the mRNAs encoding

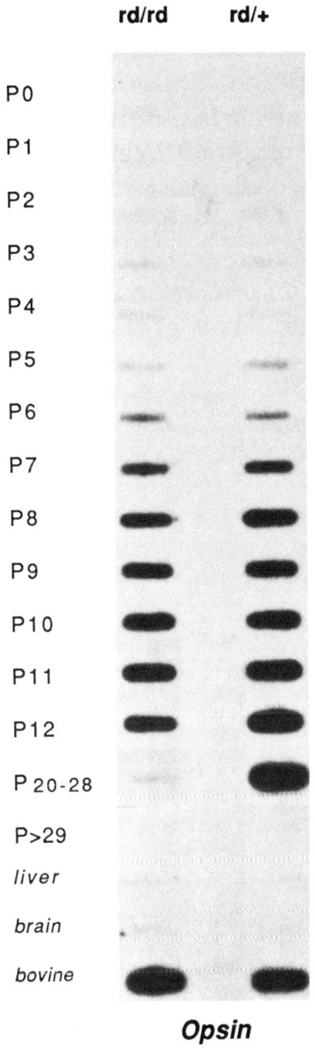

Figure 2. Hybridization of opsin cDNA to a slot blot of RNAs obtained at different times during postnatal development of *rd*/*rd* and *rd*/+ mouse retinas. (Age in postnatal days is indicated on the left).

the β subunit of transducin and the 48-kDa protein, we found that they are already expressed at low levels at birth in both *rd* and control retinas and remain fairly low until day 6 when their concentration begins to increase. A hybridization signal for the γ subunit of transducin can be observed by postnatal day 6 and for the α subunit of transducin by postnatal day 7 in both types of retinas. [It is possible that the appearance of the latter mRNA at a later time in the development of the mouse retina is related to the sudden increase in cyclic GMP-phosphodiesterase activity which occurs at 8-9 postnatal days in the normal retina (Farber and Lolley, 1976)]. Thus, the onset of expression for each mRNA studied is the same in normal and in diseased retinas. In addition, as observed with opsin mRNA, the concentrations of the mRNAs for 48-kDa protein and the α and γ subunits of transducin are comparable in normal and *rd* retinas from the time of their first appearance to postnatal days 11 or 12. At this

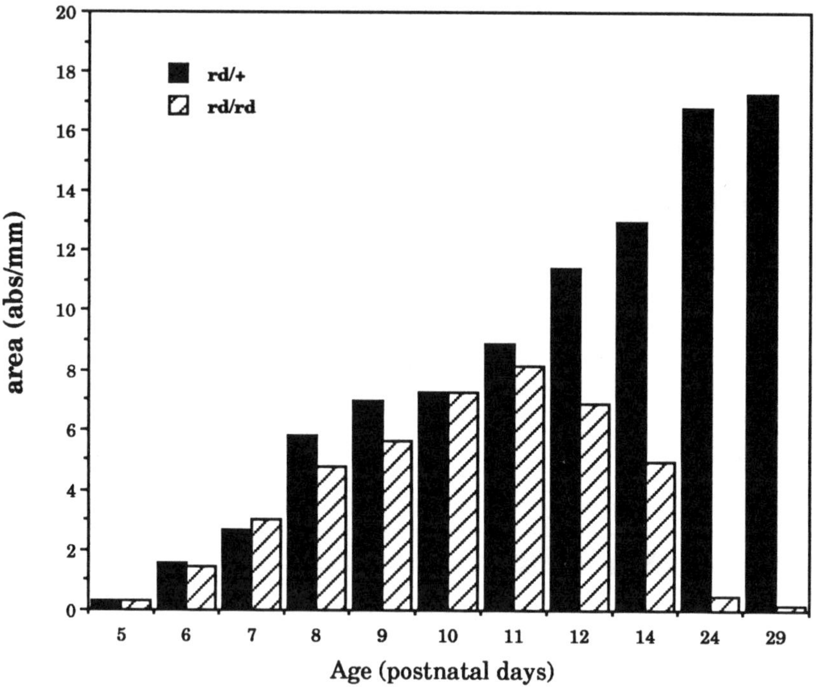

Figure 3. Developmental expression of opsin mRNA in *rd* / *rd* and *rd* / + mouse retinas. Each bar in the histogram represents the mean of the areas obtained by densitometric scanning of autoradiograms (such as that in Figure 2) of three RNA slot blots hybridized to labeled opsin cDNA.

age all the messages begin to decrease in the *rd* retina relative to the control.

The onset of expression of the proteins translated from the mRNAs that we had studied was determined in a collaborative effort with Dr. Theo van Veen (Bowes et al., 1988). This was done using immunocytochemical methods. The findings of this study correlated precisely, in terms of when the proteins are first detected, with the results that we had obtained with the mRNAs. Furthermore, we determined that each protein appears at the same time in both developing normal and *rd* retinas. From these results we concluded that a delay in expression of any of these five proteins involved in the activation/deactivation of cyclic GMP-phosphodiesterase does not seem to be associated with the cause of the *rd* disease. However, a difference in the distribution of the same proteins within the photoreceptor cells of normal and mutant retinas became apparent from these studies. After postnatal day 7, opsin and transducin are localized throughout the visual cell in the *rd* retinas but they are found only in the outer segments of control retinas. Immunoreactivity to the 48-kDa protein is observed in all areas of the visual cells in both *rd* and normal retinas during development. The 48-kDa protein is the only one of the five proteins analyzed that remains at 2 months of age in the *rd* retina although it is no longer evident in the 6-month-old *rd* mouse.

As a first approach to finding whether any modification in the composition of cyclic GMP-phosphodiesterase had occurred at the molecular level, we decided to clone the different subunits of the enzyme from normal and *rd* cDNA retinal libraries. We started with the smallest, the γ subunit, and isolated a cDNA of 482 bp from a +/+ retinal cDNA library which contains the complete coding moiety of the gene (261 bp), 121 bp of the 5'-untranslated and 100 bp of the 3'-untranslated regions (Tuteja and Farber, 1988). This cDNA was used as a probe for the detection of the mRNAs encoding the γ subunit of the enzyme from normal and *rd* retinas. On Northern blots, the corresponding messages had the same size (approximately 900 bp) and intensity, suggesting that there are no major differences between them. In addition, when slot blots prepared with RNA obtained from mouse retinas at each day of postnatal development were hybridized with the cDNA for the γ subunit of cyclic GMP-phosphodiesterase, we could not find any difference in the amount or time of onset of expression of the corresponding normal and *rd* mRNAs (Figure 4).

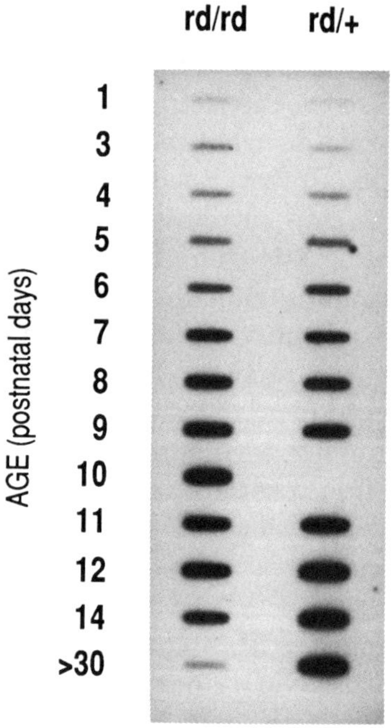

Figure 4. Developmental expression of the γ subunit of cyclic GMP-phosphodiesterase mRNA in *rd*/*rd* and *rd*/+ mouse retinas. The autoradiogram shows the hybridization of the cDNA encoding the γ subunit of cyclic GMP-phosphodiesterase to a slot blot of RNAs obtained from retinas of *rd*/*rd* and *rd*/+ mice at different times during postnatal life.

That the γ subunit of the *rd* enzyme is normal at the molecular level was confirmed by the cloning and sequencing of its corresponding cDNA (Tuteja et al., 1989). This cDNA has 926 bp including a poly (A^+) tail of 50 bp at the 3' end, 261 bp of coding region, 121 bp of 5'-untranslated and 544 bp of 3'-untranslated regions. Figure 5 shows that the *rd* coding region has only one base changed from that of the normal mouse coding sequence. This does not affect the deduced amino acid sequence of the γ subunit of the *rd* enzyme, which has 100% identity with the normal enzyme. Therefore, this subunit of cyclic GMP-phosphodiesterase does not seem to be defective in the *rd* mouse retina.

The location of the *rd* gene on mouse chromosome 5 has been established for many years (Sidman and Green, 1965). Prior to beginning the laborius process of cloning and sequencing

rd/rd	ATG AAC CTG GAG CCA CCC AAG GGT GAG ATT CGG TCA GCC ACC CGG GTG ATA GGA
+/+	ATG AAC CTG GAG CCA CCC AAG GGT GAG ATT CGG TCA GCC ACC CGG GTG ATA GGA
	GGA CCA GTC ACC CCC AGG AAA GGA CCA CCT AAG TTT AAG CAG CGG CAA ACA AGG
	GGA CCA GTC ACC CCC AGG AAA GGA CCA CCT AAG TTT AAG CAG CGG CAA ACC AGG
	CAG TTC AAG AGC AAG CCC CCC AAG AAA GGC GTG CAA GGG TTT GGG GAT GAC ATC
	CAG TTC AAG AGC AAG CCC CCC AAG AAA GGC GTG CAA GGG TTT GGG GAT GAC ATC
	CCT GGA ATG GAA GGC CTG GGG ACA GAT ATC ACC GTC ATC TGC CCT TGG GAG GCC
	CCT GGA ATG GAA GGC CTG GGG ACA GAT ATC ACC GTC ATC TGC CCT TGG GAG GCC
	TTC AAT CAC CTA GAG CTG CAC GAG CTG GCC CAG TAT GGC ATT ATC TAG
	TTC AAT CAC CTA GAG CTG CAC GAG CTG GCC CAG TAT GGC ATT ATC TAG

Figure 5. Nucleotide sequences in the coding region of the cDNAs for the γ subunit of cyclic GMP-phosphodiesterase from *rd* / *rd* and normal mouse retinas. The only nucleotide difference between the two sequences is indicated by the arrow.

the long α and β subunits of cyclic GMP-phosphodiesterase and also to determine whether it was worth the effort of carrying out similar work for the proteins involved in the action of the enzyme, we decided to study the chromosomal location of all these visual cell components to see if any mapped to mouse chromosome 5. Only those proteins located on the chromosome containing the *rd* gene would then be studied in detail. Gene mapping results were obtained by genomic Southern analysis of somatic cell hybrids. These hybrids are formed by the fusion of an established line of Chinese hamster cells with mouse spleen or peritoneal cells. After fusion, the hybrids lose mouse chromosomes preferentially and in a random fashion. Thus, different hybrid clones have different mouse chromosome content. By Southern blot hybridization of the DNAs from a panel of somatic cell hybrids with a given cDNA probe, it is possible to assign a mouse gene or related sequences to a particular mouse chromosome or to more than one chromosome. At this time we have mapped the genes for the α subunit of rod transducin to mouse chromosome 9 (Danciger et al., 1989a), the β subunit of transducin to mouse chromosome 4 (Danciger et al., 1990a), the α subunit of cyclic GMP-phosphodiesterase to mouse chromosome 18 (Danciger et al., 1990b), the γ subunit of cyclic GMP-phosphodiesterase to mouse chromosome 11 (Danciger et al., 1989b), 48-kDa protein to mouse chromosome 1 (Danciger et al., 1989c), and rhodopsin to mouse chromosome 6. Unfortunately, none of the available cDNA probes that we have investigated maps to mouse chromosome 5. This indicates that none of the corresponding proteins are the primary site of the *rd* disease.

Several years ago, concomitant with the work described above, we devised a different approach to the search for the primary cause of the *rd* mouse disease, just in case we would not be able to identify it by the conventional methods that we were using. Our strategy was based on the assumption that the mRNA encoded by the *rd* gene would be expressed differently (either be absent or manifested as an increased or decreased amount of mRNA) in affected *rd* photoreceptors compared to normal visual cells.

We took advantage of the fact that the adult *rd* retina is completely devoid of photoreceptor cells, to design a procedure using subtractive cloning to isolate a pool of photoreceptor-specific cDNAs (Bowes et al., 1989). The first step in this procedure was the preparation of a radiolabeled probe consisting of photoreceptor-enriched cDNAs. This probe was obtained by

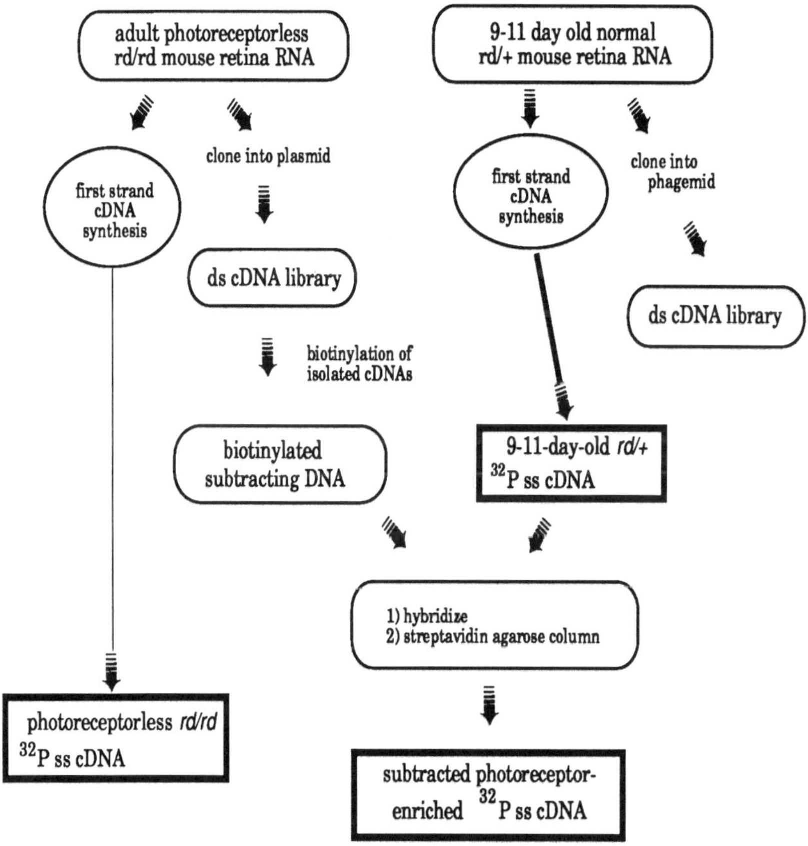

Figure 6. Outline of strategy used for obtaining the different probes (heavy boxes) used for the isolation of the candidate *rd* cDNA.

hybridizing a large excess of biotinylated, denatured DNA derived from an adult *rd/rd* cDNA library with radiolabeled single stranded cDNAs of 9-11-day-old normal *rd/+* mouse retina. In all these studies we used 9-11-day-old affected *rd/rd* mice and their morphologically normal, heterozygous *rd/+* littermates in order to obtain the greatest amount of age-matched retinal tissue before full degeneration of the *rd/rd* visual cells had occurred. Following hybridization, the biotinylated driver DNA and the non-photoreceptor *rd/+* cDNAs hybridized to it were separated from the unhybridized, *rd/+* cDNAs on a streptavidin-Sepharose column. The resulting photoreceptor-enriched probe (single stranded, unhybridized *rd/+* cDNAs) was then used to screen a 9-11-day-old normal *rd/+* retinal cDNA library (approximately 100,000 plaques) for homologous sequences. 588 clones were identified and isolated. These putative photoreceptor-specific clones were rescreened by differential hybridization with adult *rd/rd* single strand cDNAs, and 400 of them were confirmed as photoreceptor-specific.

The selection of candidate cDNAs for the *rd* gene was carried out by an additional use of differential hybridization. This time we screened the photoreceptor-specific clones with single stranded cDNAs from 9-11-day-old *rd/+* and *rd/rd* mouse retinas. We found 3 clones that hybridized differently with the two probes. However, when the DNAs of the 3 clones were used to probe Northern blots of normal and affected mouse retinal RNAs, only one clone, zr.408, showed a difference: it hybridized more faintly to the *rd/rd* RNA than to the age-matched normal RNA. The other 2 photoreceptor-specific clones hybridized with the same intensity to the *rd/rd* and *rd/+* RNAs.

The zr.408 cDNA has the following properties which, in our judgement, make it a strong candidate for the *rd* gene:

1) Consistently, on different Northern blots prepared with RNA extracted from different sets of retinas, the putative *rd* cDNA hybridizes a 3.3 kb mRNA from normal tissue and, only very weakly, a larger 3.6 kb mRNA from *rd/rd* retina (Figure 7).

This size difference could result from alternative splicing, polyadenylation site differences, or the presence of a structural alteration in the corresponding gene of the *rd*/*rd* mouse.

2) The mRNA hybridized by the candidate *rd* cDNA is detectable on postnatal day 1 in both normal and diseased

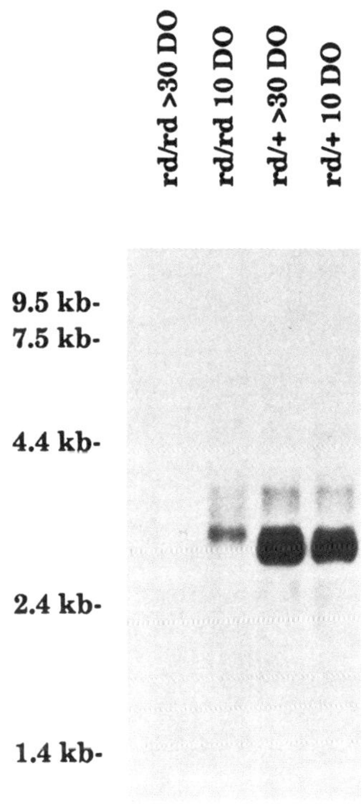

Figure 7. Northern blot analysis of retinal mRNAs isolated from *rd*/*rd* and *rd*/+ mice at different postnatal ages and hybridized with the cDNA for the candidate *rd* gene. Note the absence of message in the retinas of *rd*/*rd* mice older than 30 days.

A.

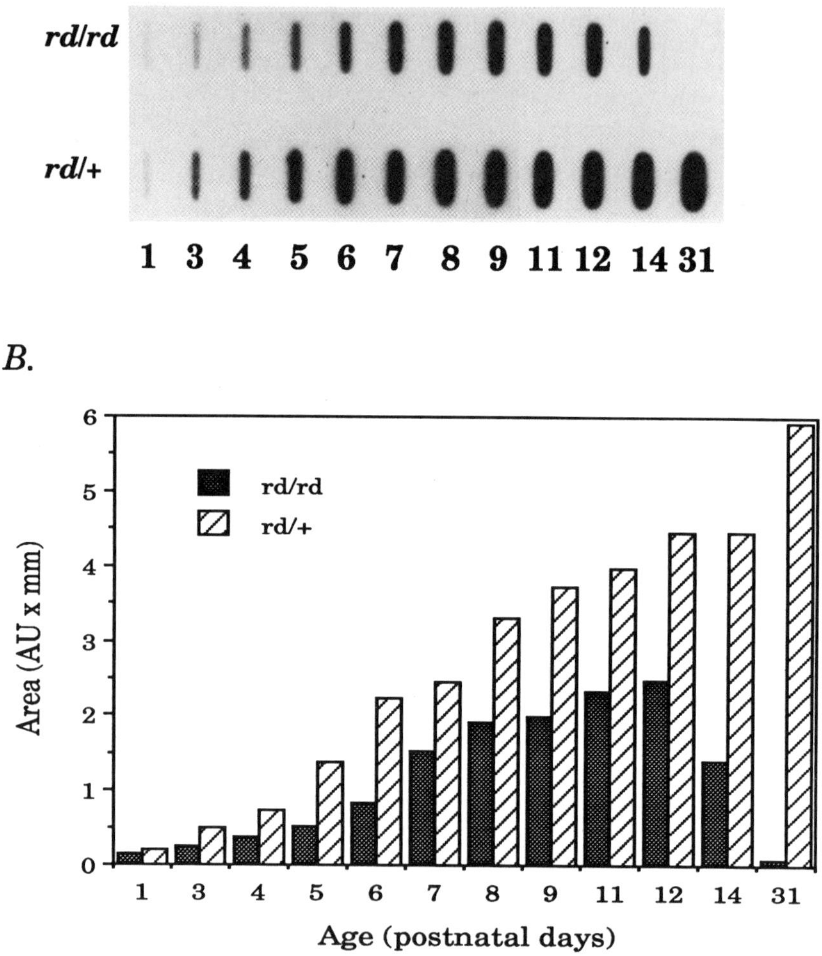

Figure 8. Developmental expression of mRNA hybridized by candidate *rd* cDNA. A) Autoradiogram of slot blot of RNA isolated from *rd*/*rd* and *rd*/+ mouse retinas at different times during postnatal development. B) Histogram of results from densitometric scanning of autoradiogram in (A). Note signal is detectable by postnatal day 1.

retinas and its concentration increases steadily during development but is always lower in the *rd/rd* retina (Figure 8). Furthermore, just as we observed in the *rd* retina with all the other probes described before, there is a sharp decrease in the levels of zr.408 mRNA at 14 postnatal days and no detectable mRNA signal above background at 31 postnatal days. Thus, the expression of the putative *rd* cDNA is abnormal in the *rd/rd* retina several days before cyclic GMP levels become elevated in the visual cells. This constitutes the earliest molecular lesion demonstrated to date in the *rd* mouse disease.

3) The candidate *rd* cDNA hybridizes different sized restriction fragments from genomic normal (+/+) and *rd/rd* DNAs, suggesting a structural difference between the +/+ and *rd/rd* gene sequences at sites homologous to zr.408. Furthermore, both the +/+ and *rd/rd* polymorphisms are present in all cases in the *rd*/+ DNA at about half the intensity observed for each in the DNAs from the homozygous mice, confirming the differences between the *rd/rd* and +/+ DNAs.

Acknowledgements

The authors would like to thank Dr. John Flannery for critical comments on the manuscript. This work was supported by USPHS Research grants EY02651 and EY08285 and by grants from the National Retinitis Pigmentosa and the George Gund Foundations.

References

Aguirre G, Farber DB, Lolley RN, Fletcher RT and Chader GJ (1978). Rod-cone dysplasia in Irish Setters: a defect in cyclic GMP metabolism in visual cells. Science 201:1133-1134.

Baehr W, Devlin MJ and Applebury ML (1979). Isolation and characterization of cGMP phosphodiesterase from bovine rod outer segments. J. Biol. Chem. 254:11669-11677.

Bowes C and Farber DB (1987). mRNAs coding for proteins of the cGMP cascade in the degenerative retina of the *rd* mouse. Exp Eye Res 45: 467-480.

Bowes C, van Veen T, and Farber DB (1988). Opsin, G-protein and 48-kDa protein in normal and *rd* mouse retinas: developmental expression of mRNAs and proteins and light/dark cycling of mRNAs. Exp Eye Res 47: 369-390.

Bowes C, Danciger M, Kozak CA and Farber DB (1989). Isolation of a candidate cDNA for the gene causing retinal degeneration in the *rd* mouse. Proc Natl Acad Sci USA 86:9722-9726.

Carter-Dawson LD, LaVail MM, and Sidman RL (1978). Differential effect of the *rd* mutation on rods and cones in the mouse retina. Invest Ophthalmol Vis Sci 7: 489-498.

Danciger M, Kozak CA and Farber DB (1989*a*). The gene for the α subunit of retinal rod transducin is on mouse chromosome 9. Genomics 4: 215-217.

Danciger M, Tuteja N, Kozak CA and Farber DB (1989*b*). The gene for the γ - subunit of retinal cGMP-phosphodiesterase is on mouse chromosome 11. Exp Eye Res. 48:303-308.

Danciger M, Kozak CA, Tsuda M, Shinohara T and Farber DB, (1989*c*). The gene for retinal S-antigen (48-kDa protein) maps to the centromeric portion of mouse chromosome 1 near *Idh-1*. Genomics 5:378-381.

Danciger M, Farber DB, Peyser, M. and Kozak CA (1990*a*). Mapping of the gene for the β-subunit of retinal transducin (*Gnb-1*) to distal mouse chromosome 4 and demonstration of additional homologous sequences on chromosomes 5 and 8. Genomics (in press).

Danciger M, Kozak CA, Li T, Applebury ML and Farber DB (1990b). Genetic mapping demonstrates that the α-subunit of retinal cGMP-phosphodiesterase is not the site of the *rd* mutation. Exp Eye Res (in press).

Farber DB and Lolley RN (1974). Cyclic guanosine monophosphate: Elevations in degenerating photoreceptor cells of the C3H mouse retina. Science 186:449-451.

Farber DB and Lolley RN (1976). Enzymatic basis for cyclic GMP accumulation in degenerative photoreceptor cells of mouse retina. J Cyclic Nucleotide Res 2: 139-148.

Farber DB and Lolley RN (1977). Light-induced reduction in cyclic GMP of retinal photoreceptor cells *in vivo*: abnormalities in the degenerative diseases of RCS rats and *rd* mice. J Neurochem 28: 1089-1095.

Farber DB, Flannery JG, Bird AC, Shuster TA and Bok D (1987). Histopathological and biochemical studies on donor eyes affected with retinitis pigmentosa. In Hollyfield JG, Anderson RE, and LaVail MM. (eds): "Degenerative Retinal Disorders", New York: Alan R. Liss, Vol. 247, pp. 53-67.

Farber DB, Park S and Yamashita C (1988). Cyclic GMP-phosphodiesterase of *rd* retina: biosynthesis and content. Exp Eye Res 46:363-374.

Fung BK-K, Hurley JB and Stryer L (1981). Flow of information in the light-triggered cyclic nucleotide casade of vision. Proc Natl Acad Sci USA 78: 152-156.

Harpold MM, Evans RM, Salditt-Georgieff M and Darnell JE (1979). Production of mRNA in Chinese Hamster cells: relationship of the rate of synthesis to the cytoplasmic concentration of nine specific mRNA sequences. *Cell* 17: 1025-1035.

LaVail MM and Sidman RL (1974). Retinal degeneration in the mouse. Arch.Ophthalmol. 91: 394-400.

Lasansky A and DeRobertis E (1960). Submicroscopic analysis of the genetic dystrophy of visual cells in C3H mice. J Biophys Biochem Cytol 7:679-684.

Lolley RN and Farber DB (1980). Cyclic GMP metabolic defects in inherited disorders of *rd* mice and RCS rats. In

Bazan NG and Lolley RN (eds): "Neurochemistry of the Retina.", New York, Oxford: Pergamon Press, pp. 427-441.

Palczewski K, Farber DB and Hargrave PA (1990). Elevated level of protein phosphatase 2A activity in retinas of *rd* mice. Invest Ophthalmol Vis Sci (submitted).

Pfister C, Kuhn H and Chabre M (1983). Interaction between photoexcited rhodopsin and peripheral enzymes in frog retinal rods: Influence on the post metarhodopsin II decay and phosphorylation rate of rhodopsin. Eur J Biochem 136: 489-499.

Shuster TA and Farber DB (1984). Phosphorylation in sealed rod outer segments: effects of cyclic nucleotides. Biochemistry 23: 515-521.

Shuster TA and Farber DB (1986). Rhodopsin phosphorylation in developing normal and degenerative mouse retinas. Invest Ophthalmol Vis Sci 27: 264-268.

Sidman RL and Green MC (1965). Retinal degeneration in the mouse; location of the *rd* locus in linkage group XVII. J Hered 56: 23-29.

Sonohara O and Shiose Y (1968). Electron microscopic study of the visual cell of inherited retinal dystrophic mice. Folia Ophthalmol Jpn 19: 77-86.

Tuteja N and Farber DB (1988). γ-subunit of mouse retinal cyclic GMP phosphodiesterase: cDNA and corresponding amino acid sequence. FEBS Lett 232:182-186.

Tuteja N, Tuteja R and Farber DB (1989). Cloning and sequencing of the γ-subunit of retinal cyclic GMP-phosphodiesterase from *rd* mouse. Exp. Eye Res. 48: 863-872.

Woodford B, Liu Y, Fletcher R, Chader G, Farber D, Santos-Anderson R and T'so M (1982). Cyclic nucleotide metabolism in inherited retinopathy in Collies: A biochemical and histochemical study. Exp Eye Res 34: 703-714.

Molecular Biology of the Retina: Basic and
Clinically Relevant Studies, pages 87–114

MOLECULAR CHARACTERIZATION OF THE RETINAL DEGENERATION SLOW (*rds*) MUTATION IN MOUSE

Gabriel H. Travis

Department of Psychiatry
University of Texas Southwestern
Medical Center
Dallas, Texas 75235

INTRODUCTION

The decision to study retina is an easy one to justify. As humans we are visual animals; most of what we know about the world we have learned by seeing. It seems entirely fitting that we understand the processes involved in sight. Even to a neurobiologist with an unbiased interest in all central neural systems, the retina is an attractive object of study. It is a repetitive structure whose simplicity approaches that of invertebrate systems, yet it is developmentally part of the brain. Input into the retina can be readily manipulated experimentally by light stimulation, and the output of the retina is accessible and neatly bundled as the optic nerve. Finally, a great deal is already known about the biology of the retina.

What, however, could be the justification for studying mutations which pervert the normal workings of this beautiful formation? One of the most potent strategies for studying the mammalian brain has been to lesion a structure of interest, such as a fiber tract or a nucleus, and then to characterize the resulting changes in the whole system. These observed changes tell us something about the function of the ablated structure and about the entire system as well. This strategy is equally valid for studying a neural system, such as the retina, on the molecular level. Unfortunately, we are not yet able to create lesions in genes at will, although this technology is rapidly advancing. We can however, take advantage of genetic lesions that occurred spontaneously in nature. By cloning the gene for these mutations and then characterizing the gene products, we have the opportunity to make important structure-function correlations about retinal proteins.

In this chapter we will consider one mutation in mouse with a phenotype in retina. The first section will be a review of several important studies of this mutation that opened the door to cloning the gene. The remainder of the chapter will be a description of experiments performed by myself and my collaborators that lead to cloning the gene and to a beginning characterization of the gene product.

Retinal degeneration slow (*rds*) is one of four mouse mutants that manifest degeneration of photoreceptors in the retina with preservation of all other retina cell types. The other mutations are retinal degeneration (*rd*) (Sidman and Green, 1965), Purkinje cell degeneration (*pcd*) (Mullen *et al*, 1976), and nervous (*nr*) (Sidman and Green, 1970). The *rds* mutation was first discovered by Van Nie *et al* (1978) when he observed blindness in adult 020/A mice. Retinal degeneration slow was so named because the rate of photoreceptor death in this mutation was much slower than in the previously described *rd* mouse. At first there was some confusion between the two mutants. To prove that they were different, crosses between mice with the genotypes *rds/rds* and *rd/rd* were made. These gave offspring with essentially normally retina. Since both mutations are recessive, this proved that *rd* and *rds* are nonallelic. The *rds* locus was later mapped to chromosome 17 near the H-2 complex (Demant *et al*, 1979). Mouse strain 020/A was produced in the Netherlands and is unrelated to any other commonly used mouse strain (Van Gulik and Kortweg, 1940). An exhaustive survey of retinal histology in different mouse strains by Sidman and Green (1965) revealed none with a phenotype similar to 020/A. This suggests that the *rds* mutation arose spontaneously only once.

MORPHOLOGY

Phenotypically, *rds/rds* homozygotes are characterized by a biphasic pattern of abnormal photoreceptor development, followed by slow photoreceptor degeneration. The retina undergoes entirely normal development and differentiation of cell types until the time of appearance of photoreceptor outer segments. While the other retinal cells continue their normal development, the photoreceptors fail to elaborate outer segments (Sanyal *et al*, 1980; Sanyal and Jansen, 1981). Of course, photoreceptor disc structures are completely absent. The photoreceptor inner segments, however, including the ciliary projections, are normal (Cohen, 1983). The synaptic termini of photoreceptors with second-order retinal neurons also appear morphologically normal.

The process of photoreceptor degeneration in *rds/rds* mice is first detectable histologically during the third postnatal week as a reduction in the thickness of the outer nuclear layer (ONL). The rate of loss of photoreceptors is greatest during the first few weeks and then becomes more gradual. Approximately half the photoreceptors are gone by three months and virtually all are gone by one year in 020/A (Sanyal *et al*, 1985). Electron microscopic analysis of degenerating photoreceptors showed non-specific changes including vacuolated and fragmented inner segments, disorganized perikaryal cytoplasm, pycnotic nuclei, and swollen and vacuolated terminals (Jansen and Sanyal, 1984). Rods and cones are affected approximately equally during the early phase of degeneration. During the later phase, rods degenerate at a slightly faster rate than cones. Photoreceptors situated in the periphery and the center of the retina appear to degenerate at approximately equal rates (Sanyal *et al*, 1985). The pigment epithelium shows accumulation of various inclusion bodies, presumably representing phagocytic removal of photoreceptor debris (Sanyal *et al*, 1985). This effect is probably secondary to the process of photoreceptor degeneration. There are no reports of structural abnormalities in brain or any other non-retinal structures in *rds/rds* mice. These careful descriptions of the pathologic changes in *rds/rds* accurately defined the morphologic phenotype but provided no information about the underlying biochemical defect.

Although the *rds* mutation was originally described as autosomal recessive, *rds/+* heterozygotes are phenotypically somewhat abnormal. In contrast to homozygotes, the heterozygotes do form outer segments. These develop more slowly however, are reduced in length, and display irregular orientation with misalignment of outer segment discs into "whorl structures" (Hawkins *et al*, 1985). Reduced ONL thickness caused by very slow degeneration of photoreceptors is also seen in heterozygotes. This process of degeneration never reaches completion. Even in very old *rds/+* mice, visual cells are still present. Thus, in an attenuated form, the heterozygote exhibits both the dysplastic and the degenerative features of the homozygote's phenotype. As we will discuss later, the presence of a phenotype in the *rds/+* heterozygote has interesting biochemical and genetic ramifications.

BIOCHEMISTRY

The *rds* retina has been partially characterized biochemically. The biochemical defect underlying the mutant phenotype is not known. Given the absence of photoreceptor outer segments and outer segment discs, reduced levels of the rhodopsin protein would be predicted in *rds/rds* retina prior to degeneration. The level of rhodopsin was measured in pre-degenerate homozygotes immunologically with anti-bovine rhodopsin antisera and was found to be 3% of normal (Schalken *et al*, 1985). We measured steady state levels of the opsin mRNA in two month *rds/rds* retina by quantitative Northern blotting. Interestingly, after correcting for partial photoreceptor loss in two month homozygotes, we found approximately normal levels of the mRNA (Travis *et al*, 1989). This implies that either the rate of translation or the stability of the opsin protein within *rds/rds* photoreceptor cells is greatly reduced. Another explanation is that the distal inner segments are rapidly packaging and exporting opsin protein in vesicles that are then being phagocytosed and degraded by cells in the pigment epithelium.

There are two lines of evidence suggesting that the visual transduction cascade is intact in pre-degenerate *rds/rds* photoreceptors, despite the absence of outer segments. Firstly, electroretinographic (ERG) recordings were made from *rds/rds* mice (Reuter and Sanyal, 1984). As would be expected, in one year old animals, the ERGs were completely absent. This was in keeping with complete loss of photoreceptors by that age. In one month animals however, the ERG recordings showed reduced amplitude of a-waves and b-waves, but with normal time-to-peak intervals. This implies that all of the steps in visual transduction, from rhodopsin activation through to synaptic transmission to the second order neurons, is intact. This is puzzling since several of these process are thought to occur on the optic discs within the outer segments, which are absent in *rds/rds* retina. It has been shown however, that rhodopsin is present in plasma membrane as well as the optic discs in normal photoreceptors (Jan and Revel, 1974; Papermaster *et al*, 1978). Thus, phototransduction may occur exclusively on the plasma membrane within the inner segments of *rds/rds* photoreceptors.

The response to light of photoreceptors in *rds* has been tested by direct biochemical measurement. Cohen (1983) observed approximately two-fold reduction in cGMP levels in the retinas of light-adapted *rds/rds* mice compared to dark-adapted mice at 21 days. A similar effect was seen at 163 days, an age where there is very significant loss of photoreceptors in *rds/rds* mice. This reduction in cGMP levels with light-adaptation is comparable to the reduction seen in wild-type albino BALB/c mice. This is further evidence that the visual transduction cascade, at least through the activation of cGMP-phosphodiesterase, is intact in *rds/rds* retina.

The absolute levels of cGMP in *rds/rds* retina were reduced approximately 2-fold at 21 days and approximately eight-fold at 163 days compared to a wild-type albino strain (BALB/c) (Cohen, 1983). Levels of cGMP-phosphodiesterase activity showed a similar pattern of less than two-fold reduction compared to BALB/c at 20 days, and nearly five-fold reduction compared to wild-type at 60 days (Sanyal *et al*, 1984). These changes in the absolute levels of cGMP and cGMP-phosphodiesterase activity are probably a secondary effect of non-development of the outer segments, superimposed upon slow degeneration of the photoreceptors in these mice.

EXPRESSION OF THE *rds* GENE

As we have discussed, the morphological abnormalities of the *rds* mutation are confined to photoreceptor cells. However, this does not say anything about the site of expression of the *rds* gene. Before any strategies for gene cloning can be conceived, it is critical to determine in what cells the *rds* gene is expressed in wild-type mice. In an elegant experiment, Sanyal and Zeilmaker (1984) prepared chimeric (tetraparental) mice by aggregating two different blastocyst-stage embryos with the genotypes *rds/rds* and +/+. The offspring of these manipulations were chimeras with patchy distribution of the two genotypes in all their tissues. This same patchiness was present in the retina. Here, retinal degeneration occurred in discrete areas with adjacent patches of entirely normal retina. Since *rds/rds* was on an albino background, while +/+ was on an agouti background, the genotype of patches in the pigment epithelium could readily be determined. No correlation was seen between areas of retinal degeneration and the genotype of the overlying pigment epithelium. A similar experiment was performed using *rds*/+ ↔ +/+ aggregation chimeras (Sanyal *et al*, 1986). The results were similar: patchy dysplastic and degenerative changes of photoreceptors were seen with no relation to the genotype of the overlying pigment epithelium.

The results of these experiments suggested that the *rds* gene is probably cell autonomous, which is to say that it is expressed normally in photoreceptors. Since no correlation between the genotype of the degenerating retina and the genotype of the pigment epithelium was observed, the *rds* phenotype is probably not mediated through the pigment epithelium. An unlikely possibility that cannot be ruled-out by these experiments is that the *rds* gene is normally expressed in other nearby retinal neurons, and that degeneration of photoreceptors is mediated by a trans-cellular effect.

GENE CLONING STRATEGY

The strategy we used to clone the *rds* gene (Travis *et al*, 1989) was based upon a prediction about its pattern of expression. This prediction was supported by two of the observations presented above: (1) the wild-type *rds* gene is expressed in photoreceptors, and (2) the process of abnormal development and subsequent degeneration in *rds* mutants is restricted to photoreceptors. We hence made the prediction that the wild-type *rds* gene is expressed specifically within photoreceptors in the retina, and is located on chromosome 17. The strategy that we employed involved isolating a collection of cDNA clones representing mouse photoreceptor-specific mRNAs, and then analyzing to see if any of these mRNAs were encoded by a gene on chromosome 17. We would thus consider a cDNA with these characteristics as a candidate clone of the *rds* mRNA. Confirming such a clone would involve comparing the corresponding mRNAs in wild-type and *rds/rds* retina, and comparing the structure of the corresponding wild-type and mutant genes.

ISOLATION OF PHOTORECEPTOR-SPECIFIC mRNA CLONES

We took advantage of the unrelated mutant, *rd/rd* in order to isolate photoreceptor-specific mRNAs. Mice with this mutation manifest normal retinal development followed by rapid degeneration of the photoreceptors beginning by day 10 (Sidman and Green, 1965; LaVail and Sidman, 1974). The photoreceptors are completely absent by 4 weeks, while all other retinal cells are spared (Carter-Dawson *et al*, 1978). Thus, an mRNA present in wild-type (C57BL/6) but absent from *rd/rd* (C3H/HeJ) retina is photoreceptor specific.

We employed a subtractive cDNA cloning procedure that involves the use of cloned hybridization-driver, and takes advantage of the enhanced hybridization kinetics seen in a phenol emulsion (Travis and Sutcliffe, 1988). Poly(A)$^+$ RNA was prepared from the retinas of 6-7 week wild-type (C57BL/6) and 6-7 week *rd/rd* (C3H/HeJ) mice. First, retinal cDNA libraries were prepared from these two RNA samples. A ^{32}P-labelled cDNA probe was also prepared from the C57BL/6 RNA. This probe was hybridized to an excess of unlabelled C3H/HeJ mouse retina cDNA in a phenol emulsion. Sequences in common between C5BL/6 and C3H/HeJ (non-photoreceptor-specific mRNAs) will form hybrids under these conditions. The fraction of the probe that remained single-stranded during the reaction was thus enriched with mRNAs not in common between C5BL/6 and C3H/HeJ (photoreceptor-specific mRNAs). This single stranded, ^{32}P-labelled DNA was separated from the mixture by chromatography on hydroxyapatite. The resulting probe was used to screen approximately 10^5 colonies of the cDNA library made from C57BL/6 mouse retina cDNA. Clones that gave signals on the autoradiograph with the subtracted probe were placed in a grid and re-screened by differential colony hybridization with unsubtracted, ^{32}P-labelled cDNA from C5BL/6 and C3H/HeJ retina. We isolated seventy five clones that gave signals with unsubtracted C57BL/6 cDNA, but not with unsubtracted C3H/HeJ cDNA.

Grid-hybridization with individual clones showed that there were approximately 12 different mRNAs represented in this collection of differentially expressed cDNAs. To confirm their photoreceptor-specificity, we performed Northern blot analysis of these clones with poly(A)$^+$ RNA from C57BL/6 and C3H/HeJ retina, and C57BL/6 whole brain. The Northern blot hybridization pattern for six of these is shown in Figure 1. Interestingly, all of the cloned mRNAs that were photoreceptor-specific in retina were also undetectable in brain.

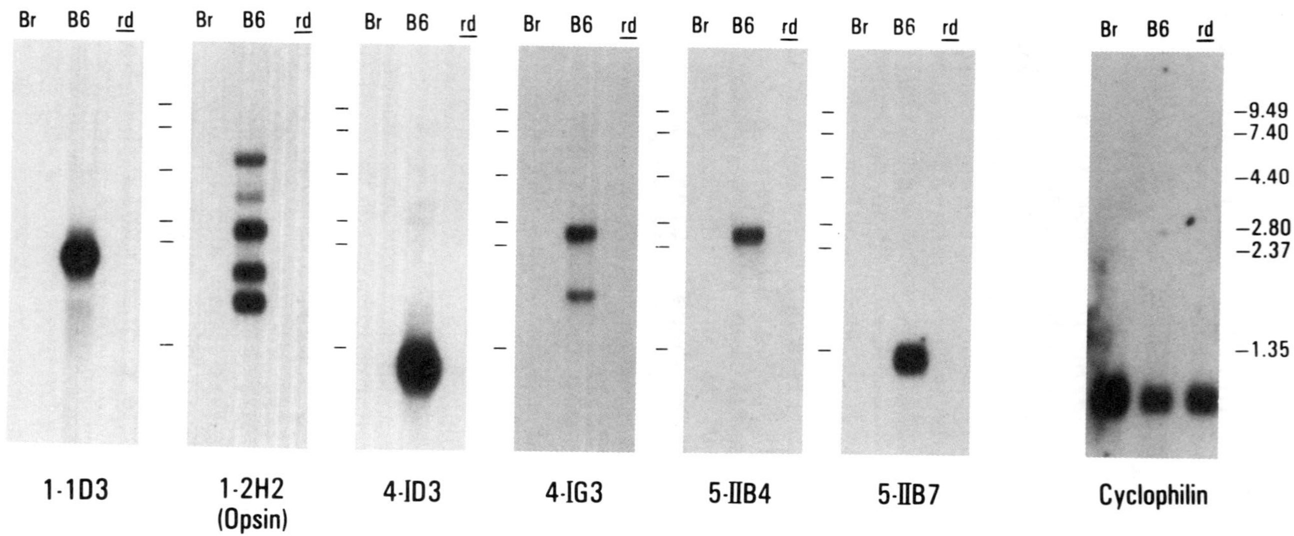

Figure 1 Northern blot hybridizations of six photoreceptor-specific mRNAs. Each lane contained 2 μg of poly(A)$^+$ RNA. Br, C57BL/6 total mouse brain; B6, C57BL/6 (+/+) retina; rd, C3H/HeJ (*rd/rd*) retina. RNA size standards are indicated to the right in kB. Probes were prepared from cDNA clones derived from the C57BL/6 minus C3H/HeJ retinal subtractive screen. Ubiquitously expressed rat cyclophilin (Danielson *et al*, 1988) was used as a control for equal loading of lanes.

CHROMOSOME ASSIGNMENT

The next step was to determine the chromosome assignment for the genes encoding each of the cloned photoreceptor-specific mRNAs. This we did by probing a panel of 16 mouse x hamster somatic cell hybrid DNAs (Hoggan *et al*, 1988) with each of the 12 different photoreceptor-specific cDNAs. These hybrid cell lines each contained a set of several different known mouse chromosomes in addition to the entire hamster genome. DNA from these cells was digested with a restriction endonuclease that had been shown in pilot experiments for each cDNA probe to permit discrimination of the mouse and hamster alleles. Clone IG3 mapped to chromosome 17 with 100% concordance (Table 1). The remaining 11 mRNA clones mapped to other autosomes.

Clone IG3 was thus a candidate for the *rds* mRNA, since it met our prediction criteria of being photoreceptor-specific and being encoded by a gene on chromosome 17. On the Northern blot (Figure 1), the 800 bp clone IG3 detected two bands of approximately 2.7 kB and 1.6 kB in wild-type retina. A full-length cDNA of the longer mRNA was isolated by probing the C57BL/6 retinal cDNA library with clone IG3. Clone IG3.B9A contained an insert of 2.7 kB.

THE IG3.B9A mRNA HAS ALTERED ABUNDANCE AND MOBILITY IN THE MUTANT

To test whether IG3.B9A was a clone of the *rds* mRNA, poly(A)$^+$ RNA was prepared from the retinas of 1-2 month (pre-degenerate) *rds/rds* (BALB/c background) and age-matched wild-type BALB/c mice. These RNA samples were analyzed by Northern blotting using the full-length cDNA isolate, IG3.B9A as a probe. Two mRNA species of approximately 1.6 kB and 2.7 kB were detected in the BALB/c retina lane. Fainter bands of higher molecular weight were also seen in the BALB/c

TABLE 1 Segregation pattern of gene for IG3.B9A

Chromosome	Hybridization pattern +/+	-/-	+/-	-/+	% Discordance
1	4	5	4	1	35.7
2	5	3	3	3	42.9
3	2	4	3	0	33.3
4	3	3	5	2	53.8
5	0	6	8	0	57.1
6	3	4	5	2	50.0
7	4	0	4	6	71.4
8	3	5	5	0	38.5
9	1	5	7	1	57.1
10	1	6	7	0	50.0
11	0	4	6	0	60.0
12	4	1	2	3	50.0
13	3	2	5	3	61.5
14	1	5	7	1	57.1
15	6	0	0	4	40.0
16	2	4	5	1	50.0
17	8	6	0	0	0.0
18	3	2	4	3	58.3
19	2	2	6	3	69.2
X	4	5	4	1	35.7

Southern blots containing DNA from 16 mouse X hamster somatic cell hybrids were probed with cDNA clone IG3. Column designations: +/+, chromosome and IG3 hybridization-signal both present; -/-, chromosome and hybridization-signal both absent; +/-, chromosome present and hybridization-signal absent; -/+, chromosome absent and hybridization-signal present.

---

lane, possibly representing unspliced mRNA precursors. None of the bands visualized in the BALB/c lane were present in the *rds/rds* retina lane. The IG3.B9A probe did detect two faint, closely spaced bands in *rds/rds* of approximately 12 kB (Figure 2). From this result it appeared very likely that IG3.B9A was a clone of the *rds* mRNA.

When the same Northern blot was reprobed with a cDNA clone of the opsin mRNA, the BALB/c and *rds/rds* retina lanes exhibited identical hybridization patterns, although the signal

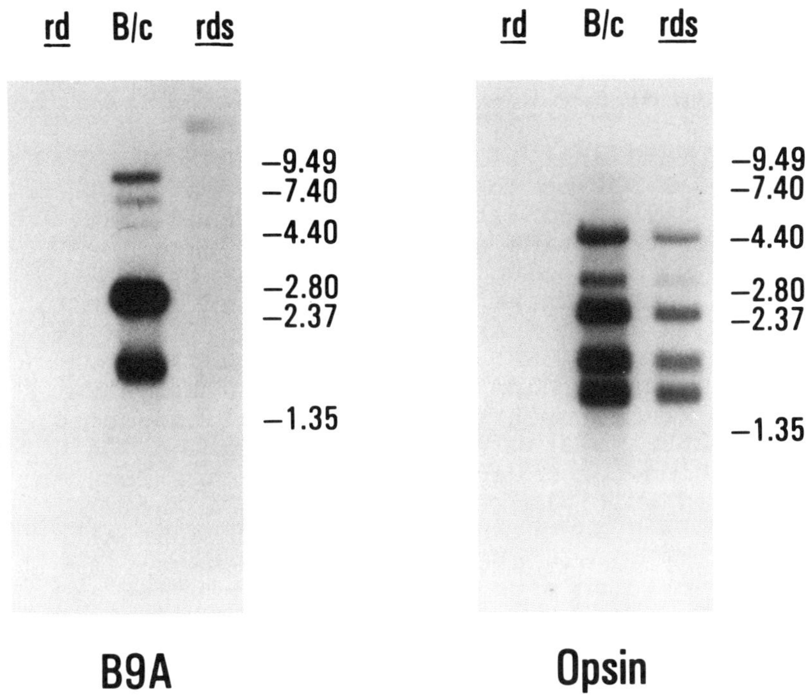

Figure 2 The IG3.B9A mRNA has aberrant size and abundance in *rds/rds* retina. Each lane contained 2 μg of poly(A)$^+$ RNA. rd, 6-7 week C3H/HeJ (*rd/rd*) retina; B/c, 4-8 week BALB/c (+/+) retina; rds, 4-8 week BALB/c (*rds/rds*) retina. RNA size standards are indicated to the right in kB. B9A Panel: Northern blot was probed with insert from cDNA clone IG3.B9A. Opsin Panel: Same blot was reprobed with insert from cDNA clone 2H2 of mouse opsin.

---

intensity in the mutant lane was slightly reduced (Figure 2). This control experiment excluded the possibility that the observed difference in RNA-hybridization patterns with the IG3.B9A probe in wild-type and *rds/rds* retina was due to premature degeneration of photoreceptors in the mutant. It also ruled out the possibility of a general abnormality of RNA processing within photoreceptors in *rds/rds*.

STRUCTURE OF THE WILD-TYPE *rds* mRNA

Sequence analysis of clone IG3.B9A revealed a clone of 2,632 bp followed by a poly(A) tail (Figure 3). A single long open reading frame began at nucleotide 213, encoding a putative protein of 346 amino acids. The putative initiator methionine was in a context (AGCAUGG) similar to the consensus sequence for translation initiation (ACCAUGG) (Kozak, 1984 and 1986). Multiple stop codons were present in all three reading frames upstream of the initiator codon. The 3' untranslated region following the TGA termination codon was 1,382 nucleotides long and contained multiple stop codons in all three frames. A frequently found variant (AUUAAA) (Birnstiel *et al*, 1985) of the consensus polyadenylation signal (AAUAAA) (Proudfoot and Brownlee, 1976) was found 20 nucleotides upstream of the poly(A) tail. We confirmed that IG3.B9a was actually full-length clone of the 2.7 kB mRNA by primer extension analysis.

Careful examination of the sequence revealed a second polyadenylation signal, CAUAAA (Birnstiel *et al*, 1985) at position 1,627. This suggested an explanation for the 1.6 kB mRNA seen on the Northern blot. The C57BL/6 retina library was re-screened in duplicate with the complete clone IG3.B9A, and a fragment of this clone derived from the extreme 3' end. A new clone (IG3.D6) of 1.6 kB was isolated that hybridized with the complete probe, but not the 3' fragment. The same 3'end fragment probe detected only the 2.7 kB RNA in BALB/c retina, and the higher band in the 12 kB doublet in *rds/rds* retina on Northern blots. Sequence analysis of IG3.D6 showed it to be identical to IG3.B9A, except that the poly(A) tail began at position 1,651, 24 bases downstream of the putative second polyadenylation signal. This confirmed that the

```
      AAGGACTCTGCAGATACGGCGGCCTAGATTAGCTCCGGCTACCGTTACTGAGTTAACGGGGATCCCAAGCTAGGGAGGCCCCAAAATGG
90    GCAACTCCCTGCAGCTTGGGCCCATGGTGCTCTTCCCTAGACCCTAGCGGTCCAGCCCCGGAGCTCACTCGGATTAGGAGTGGAAGCTGA
                                        MetAlaLeuLeuLysValLysPheAspGlnLysLysArgValLysLeuAlaGlnGly
180   ACCGTGGGAGGCTGCTGAACGCACTCGGTAAGCATGGCGCTGCTCAAAGTCAAGTTTGACCAGAAGAAGCGGGTCAAGTTGGCCCAGGGG
      LeuTrpLeuMetAsnTrpLeuSerValLeuAlaGlyIleValLeuPheSerLeuGlyLeuPheLeuLysIleGluLeuArgLysArgSer
270   CTCTGGCTTATGAACTGGCTGTCCGTGTTGGCCGGCATCGTCCTCTTCAGCTTGGGGCTGTTCTTGAAGATTGAACTTCGCAAGAGGAGC
      GluValMetAsnAsnSerGluSerHisPheValProAsnSerLeuIleGlyValGlyValLeuSerCysValPheAsnSerLeuAlaGly
360   GAAGTGATGAATAATTCTGAGAGCCACTTTGTGCCCAACTCCCTGATAGGGGTGGGGGTCCTGTCCTGTGTCTTCAACTCTCTGGCTGGG
      LysIleCysTyrAspAlaLeuAspProAlaLysTyrAlaLysTrpLysProTrpLeuLysProTyrLeuAlaValCysIlePhePheAsn
450   AAGATCTGCTATGATGCCCTGGACCCGGCCAAGTACGCCAAGTGGAAGCCCTGGCTGAAGCCGTACCTGGCTGTCTGCATCTTCTTTAAC
      ValIleLeuPheLeuValAlaLeuCysCysPheLeuLeuArgGlySerLeuGluSerThrLeuAlaTyrGlyLeuLysAsnGlyMetLys
540   GTCATCCTCTTCCTGGTGGCTCTCTGCTGCTTTCTGTTGCGGGGCTCCCTGGAGAGCACCCTGGCTTACGGACTCAAGAATGGGATGAAG
      TyrTyrArgAspThrAspThrProGlyArgCysPheMetLysLysThrIleAspMetLeuGlnIleGluPheLysCysCysGlyAsnAsn
630   TATTATCGGGATACGGACACCCCCGGCCGGTGCTTCATGAAAAAGACCATCGACATGCTCCAGATTGAGTTCAAGTGCTGTGGGAACAAC
      GlyPheArgAspTrpPheGluIleGlnTrpIleSerAsnArgTyrLeuAspPheSerSerLysGluValLysAspArgIleLysSerAsn
720   GGCTTCCGGGACTGGTTCGAGATTCAGTGGATCAGCAATCGCTACCTGGACTTCTCCTCCAAGGAGGTCAAAGATCGCATCAAGAGCAAC
      ValAspGlyArgTyrLeuValAspGlyValProPheSerCysCysAsnProSerSerProArgProCysIleGlnTyrGlnLeuThrAsn
810   GTGGATGGGCGGTACCTGGTGGACGGCGTCCCTTTCAGCTGCTGCAACCCCAGCTCCCCGCGGCCCTGTATCCAGTACCAGCTCACCAAC
      AsnSerAlaHisTyrSerTyrAspHisGlnThrGluGluLeuAsnLeuTrpLeuArgGlyCysArgAlaAlaLeuLeuAsnTyrTyrSer
900   AACTCGGCGCACTACAGCTATGACCATCAGACTGAGGAGCTCAACCTCTGGCTGCGGGGCTGCAGGGCCGCTCTGCTGAATTACTACAGC
      SerLeuMetAsnSerMetGlyValValThrLeuLeuValTrpLeuPheGluValSerIleThrAlaGlyLeuArgTyrLeuHisThrAla
990   AGCCTCATGAATTCCATGGGCGTCGTCACACTTCTCGTCTGGCTCTTTGAGGTGAGCATCACTGCCGGACTCCGCTACCTCCACACAGCG
      LeuGluSerValSerAsnProGluAspProGluCysGluSerGluGlyTrpLeuLeuGluLysSerValProGluThrTrpLysAlaPhe
1080  CTGGAGAGTGTGTCTAACCCGGAGGACCCCGAGTGTGAGAGTGAGGGCTGGCTGCTGGAGAAGAGCGTGCCCGAGACCTGGAAGGCCTTT
      LeuGluSerPheLysLysLeuGlyLysSerAsnGlnValGluAlaGluGlyAlaAspAlaGlyProAlaProGluAlaGly
1170  CTGGAGAGCTTTAAGAAGCTGGGCAAGAGCAATCAGGTGGAGGCTGAAGGTGCAGACGCAGGCCCGGCTCCAGAGGCTGGCTGATGGCCT
1260  GGGGCCTCTCCCCCTCCTCAACACTTAGTGGACTCCAGGGACTGTGGATACCCCCTTTGTCCAGCTGAAAGTCCAAATTTCCCGAGAAAG
1350  CTGGTCACCTACTGACTCTCCTTGATGTGGGCCTTGAAGTTCAGGGTCCTTAGGGCAGGTTACAAACATTTGTGAAACGGCTGCCTCCAG
1440  ATGTGAGTGACTGAACGATGCAGCAGATGGCCAGGACTGAACGCTCACAGGACTGGCCAGTCCAGGGCTCTGTCCAAGTGTGAGTCCAGA
1530  TCTCTCATAGGTGACTGGCCACACCAAGGGCCTCTCCCCTCCTCAGTAGTGTCTGCCTCTTTTAAGCTACAAGTTCTGCATCCCAACCAT
1620  CACTTGAAACATAAAAGCAAGGTGAAAATAAGAAAACTCTAAGGGCCATGTTGTTTGTTTTCCTTAATGATAGGGTTAACAGGCTGGTCA
1710  TCATTTGCTTCATGTTACACAATGGGGGAGAAGCAATGATATTTTTTTAAGAGTATTGGACTCCCTCAGAAGGTTACCTCCACATAGAAA
1800  TTCACAGATGGCACAGGAGGTAAAAACTTTCCTTTAGTAAAATCCAGTGAGTGTCCCATACCTGCCCCATTCCCAGACTCGCAGTCACGT
1890  ACTTCAGCTGGCATGTCCCAGCATCCTCCACTCACCAAGTGCTGGGTGAGCCACTGCCTGAGAAGGCCCCGGGTGAGTTTGCTGTCCTTT
1980  GTGGGAGAGATGTGGTGATCAATTAGGAATCTGCCACTTGGGGACCAAGCCCTGCCTCCTGGTGGTCTCAGATGGCCACAACAGTCTTTG
2070  TGGACTGCACTGTCCAGGAGGGCAAAGACCCCTGGGAGCTTCTGTTCTCCCTCCCACAAAGCACTATGGGAATGCTTCCGTATCCAGGAA
2160  GATTCTGGAGATAGGTTATCTTTCCGAGCTAGCGTGTGTGGAAAGTTGCTGTTTCTTCTGGGAACGTCAGGCAAACTCTCAGGGGAAGCA
2250  AAAATATGATTTTGGTGGCACCTTGTGGGATGGCACCCACAAGGACTGTTGTGGCCATCTGAGACCACCAGGAGGCAGGGCTTGGTCCCC
2340  AAGGGCTCAGCTGCTGCTGGCCTGTGCTGTGTGGTTCTAACTCTCTCAGGGGGAAGACCAGATTAGAGATGCTGGCACACCCAACGAACT
2430  ATGGCCCCAATGTTGGTTAGTATTCTGAGTCTCTCAGAGCAAGAAATGGCCCAGATGCTGAGCCAGGCATCTGTGGCGGGGACCCTCGGG
2520  GTCCTGCCGTCACCTCTCAGGAGGGCTGAGACTATGGTGAAAGATACTCTAGACCCATTTCTAACTATGCTGTCAACCATCGTATTAAGA
2610  GATATTAAAACTATATTGTATG(A)n
```

Figure 3 Nucleotide sequence of the full-length *rds* cDNA clone IG3.B9A. Nucleotide numbering is shown to the left. The single long open reading frame has the corresponding amino acid sequence translated above. The complete sequence was determined for both DNA strands. The polyadenylation signal CATAAA for the 1.6 kB mRNA (IG3.D6), and ATTAAA for the 2.7 kB mRNA (IG3.B9A) are underlined.

---

only difference between the two *rds* mRNAs was utilization of alternative polyadenylation signals. They encode identical proteins.

A computer search of several protein sequence databases with the predicted protein encoded by the *rds* mRNA revealed no significant similarities to any known protein sequences. The protein does not have an apparent amino-terminal secretion signal sequence. It does contain three uncharged regions (amino acid numbers: 16-41, 100-122, and 252-275), representing possible membrane-spanning domains. It has three potential glycosylation sites and four sets of tandem basic residues (amino acid numbers: 11-13, 46-48, 178-179 and 324-325), possibly representing proteolytic cleavage signals. The protein also contains 13 Cys residues, including three Cys-Cys pairs. These are potential sites for intra- or inter-chain disulfide linkages.

THE *rds* GENE CONTAINS AN INSERTION IN A PROTEIN-CODING EXON

To define the site of the mutation within the putative *rds* locus, we prepared five adjacent, non-overlapping restriction endonuclease fragments spanning the full-length cDNA clone IG3.B9A (Figure 4). These fragments were used individually to probe Southern blots containing restriction digests of DNA from BALB/c and *rds/rds* mice. Restriction fragment length polymorphisms (RFLPs) were detected with several different restriction endonucleases using probe B9A-II, but not with the other four cDNA clone fragment probes. This result ruled-out the possibility of a point-mutation.

We prepared two genomic libraries by ligating DNA from BALB/c and *rds/rds* mice, partially digested with restriction endonuclease Sau3A, into phage lambda. The libraries were plated and probed with DNA from clone IG3.B9A. Two lambda clones were isolated from each library. We digested the longer clones from the BALB/c (B7) and *rds/rds* (R24) libraries, along with genomic DNA from BALB/c and *rds/rds* in parallel, with

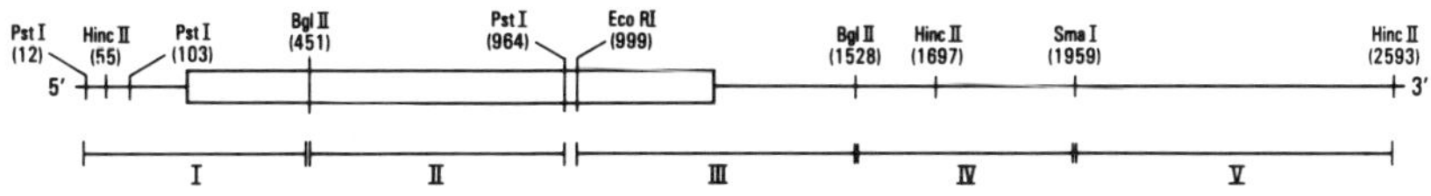

Figure 4 Restriction map of the *rds* cDNA clone IG3.B9A derived from the nucleotide sequence. The open reading frame is indicated by the open box. Restriction fragments (I-V) used as probes for Southern blot analysis are indicated below.

---

several restriction endonucleases. These DNA samples were analyzed by Southern blotting using the B9A-II fragment as a probe (Figure 5). In the BglII digests, BALB/c and *rds/rds* genomic DNA shared one band of approximately 6.5 kB. Two additional bands of approximately 4 and 4.5 kB were present only in the *rds/rds* sample. In the PstI digests, BALB/c and *rds/rds* genomic DNA shared one band of approximately 1.8 kB. Another band of 2.8 kB was present in BALB/c but absent from *rds/rds*. Three additional PstI fragments (6.3 kB, 2.6 kB (faint) and 1.1 kB) were present only in *rds/rds* genomic DNA. The hybridization pattern of the lambda DNA and the corresponding genomic DNA were virtually identical with probe B9A-II. Probe fragment B9A-III on the other hand, gave similar hybridization pattern in the mutant and wild-type lanes (data not shown). These results suggested that the mutation was the result of an insertional event, and that at least one end of the inserted DNA was present in the 6.3 kB PstI fragment detected with the B9A-II probe.

To define the mutation on the nucleotide level, the 2.8 kB PstI fragment from the wild-type and the 6.3 kB PstI fragment from the mutant detected by B9A-II were each subcloned from B7 and R24 into a plasmid vector. Restriction-

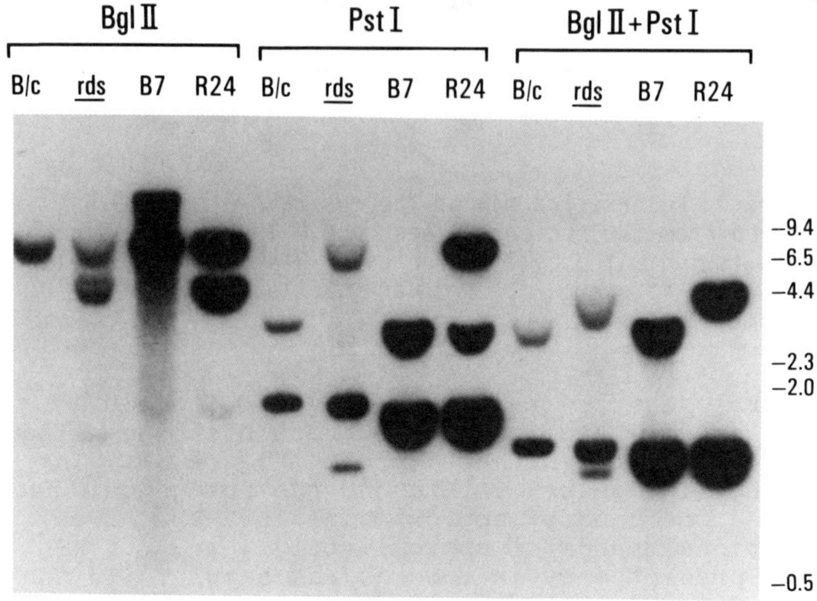

Figure 5 Southern blot analysis of genomic and lambda clone DNA with IG3.B9A fragment II probe. B/c, BALB/c DNA; rds, *rds/rds* DNA (10 μg genomic DNA per lane); B7, lambda clone B7 from BALB/c library; R24, lambda clone R24 from *rds/rds* library (0.5 ng lambda clone DNA plus 10 μg sheared salmon sperm DNA per lane). DNA samples were digested with restriction endonuclease BglII, PstI, or BglII plus PstI as indicated. DNA size standards are indicated to the right in kB.

endonuclease mapping of these subcloned PstI fragments (B:1 and R:1) revealed that the first 2.7 kB from the 5' end of each were indistinguishable. Following the third AvaI site (Figure 6), the restriction maps of B:1 and R:1 diverged. The 0.25 kB AvaI-PstI 3' fragment from the wild-type subclone B:1, and the corresponding 0.95 kB AvaI-AvaI fragment from the mutant

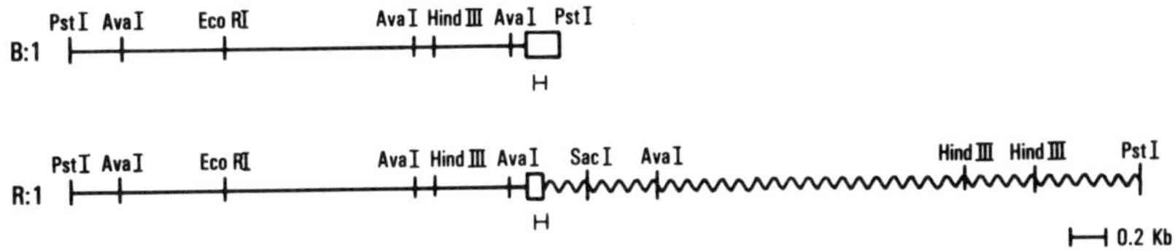

B:1 GGCCCTGTATCCAGTACCAGCTCACCAACAACTCGGCGCACTACAGCTATGACCATCAGACTGAGGAGCTCAACC

R:1 GGCCCTGTATCCAGTACCAGCTCACCAACtgccacaggcaatctgggtccctggtctgtgaagaacgggatttct

Figure 6 Restriction map of BALB/c and *rds/rds* genomic clones. B:1 is the 2.8 kB PstI genomic subclone fragment from wild-type phage lambda clone B7 detected with probe B9A-II. R:1 is the 6.3 kB PstI genomic subclone fragment from *rds/rds* phage lambda clone R24 detected with probe B9A-II. The open box indicates an exon which contains a protein-coding region. The wavy line indicates the insertion in the mutant locus. The nucleotide sequences of B:1 and R:1 in the region containing the divergence is indicated below. Capital letters in the nucleotide sequence correspond to the wild-type exon, lower-case letters correspond to the insertion. Bars underneath the restriction map show the origin of the indicated sequences.

---

subclone R:1 were isolated and their nucleotide sequences determined. The sequences were identical up to the position corresponding to nucleotide 899 in the mRNA. At this point a foreign sequence appears in R:1, disrupting the protein-coding exon (Figure 6).

Genomic sequence analysis confirmed the results of the Southern blots. The most economical explanation for the *rds* mutation on the nucleotide level is a simple insertion. Only the 5' part of the inserted element was present on subclone R:1. In an effort to identify the inserted element, we sequenced subclone R:1

entirely (data not shown). Again, no homology to any retroviral or other eukaryotic sequences was seen. Southern blot analysis was performed on genomic DNA isolated from BALB/c and *rds/rds* mice digested with restriction endonucleases PstI and BglII, using the 0.4 kB SacI-AvaI restriction fragment from the insertion of R:1 as a probe. Approximately 30 bands were detected in both the wild-type and *rds/rds* lanes, with two additional bands present in the *rds/rds* lanes that were absent from the wild-type lanes (data not shown). This suggests that the inserted element contains middle-repetitive DNA. The additional bands seen in DNA from the mutant presumably represent an extra copy of this element within the *rds* gene. The Northern blot analysis suggests that the mutant *rds* transcript is using the same polyadenylation signals as the wild-type. This means that the entire inserted element is being transcribed in mutant mice. Probing the same retina Northern blots with the 0.4 kB SacI-AvaI R:1 fragment revealed only the doublet 12 kB band in *rds/rds* (data not shown). The size of the inserted element can be estimated at 10 kB based upon the size-shift of the *rds* mRNA in the mutant.

What is the possibility that the gene identified here is not *rds*? This possibility requires that loss of function of the product of the gene that we cloned results in no phenotype, and that another gene on chromosome 17, which is also expressed in photoreceptors, bears an independent second mutation. There is another line of evidence that makes this formal possibility even less likely. The *rds* mutation originally arose on mouse strain 020/A. Somes Sanyal performed back-crossing experiments to move the *rds* mutation onto BALB/c and C3H (Sanyal, personal communication). This involved 20 crosses to BALB/c and eight crosses to C3H. Southern blot analysis using probe B9A-II was performed on the DNA from 020/A, BALB/c and C3H mice, all with the *rds/rds* genotype, and on the

DNA from the corresponding wild-type strains. All of these strains showed the identical set of RFLPs. This means that the identified gene maps to within a few centimorgans of *rds*. Thus, it seems likely that the gene we have cloned is indeed *rds*. Final proof awaits correction of the mutant phenotype by introducing a DNA construct containing the cloned wild-type gene *rds/rds* mice in a transgenic experiment.

PRELIMINARY CHARACTERIZATION OF THE *rds* GENE PRODUCT

As we have seen on the Northern blots, the *rds* gene is transcribed in the mutant. Reduced levels of the two mRNAs may be the result of either a reduced rate of transcription or reduced stability of the altered RNA. The insertion appears within the protein-coding portion of the gene at a position corresponding to residue 259. This gene encodes a protein with truncation of the C-terminal 87 amino acids, and replacement by a short, randomly encoded polypeptide before the first in-frame stop codon is encountered. There are two possible mechanisms whereby this mutation could cause the *rds* phenotype. The first possibility is that no translation product is made from the mutant *rds* mRNA, and that the phenotype arises from simple loss of function of this protein. In this case the partial phenotype in the heterozygote could be explained by haploinsufficiency; the one normal allele does not make sufficient gene product to protect the cells from slow degeneration. According to the second possible mechanism, the mutant *rds* mRNA is translated and the aberrant protein product has a directly deleterious effect on photoreceptors. Here the phenotype in the heterozygote can be explained by the presence of the abnormal protein, which is to say that the mutation is partially dominant.

As we have discussed, the *rds* protein is

novel, and hence its function is unknown. The phenotype of failed outer segment development, followed by slow degeneration of photoreceptors is not very informative about the function of the *rds* protein. We can infer that the *rds* protein is not a critical component of the cell's basic metabolic machinery, since photoreceptors can live for months in its absence. The fact that it is present exclusively within photoreceptors in the CNS suggests that its function may relate to the specialized function of the photoreceptors, namely visual transduction. However, it is unlikely to be directly coupled to the visual transduction cascade, since this pathway appears relatively intact in *rds/rds* mutants, both biochemically and by ERGs. It may therefore be a structural protein involved in the formation or maintenance of outer segment discs. In the primary structure of the protein there are three stretches of 23-25 uncharged residues, suggesting membrane-spanning domains. There is no consensus amino-terminal signal sequence, suggesting the protein is not secreted. Additionally, there are several potential glycosylation signals. An important question to resolve, in beginning an analysis of the *rds* protein, is whether it is associated with membranes.

We synthesized five non-overlapping peptides of 13 residues each, based upon the sequence predicted by the mRNA (Figure 7). Each peptide was conjugated to edestin and used to immunize rabbits. Retinal extracts were prepared from two month old wild-type (BALB/c) and *rd/rd* (C3H/HeJ) mice. Immunoblots prepared by electrophoresis of these extracts were reacted with antisera raised against each of the five peptides. Antipeptide-#4 sera detected a broad band of slightly less than 39 kD in the wild-type lane but not the *rd/rd* lane (data not shown). This reactivity was abolished by pre-incubating the antipeptide-#4 sera with #4 synthetic peptide. A plasmid was constructed in the in the vector pUR292 (Ruther and Muller-Hill, 1983) such that the coding

```
Peptide 1                                      Peptide 2
[MetAlaLeuLeuLysValLysPheAspGlnLysLysArg] [ValLysLeuAlaGlnGlyLeuTrpLeuMetAsnTrpLeu] SerValLeuAla 30
                                                             Peptide 3
GlyIleValLeuPheSerLeuGlyLeuPheLeuLysIleGluLeuArgLysArg [SerGluValMetAsnAsnSerGluSerHisPheVal   60

Pro] AsnSerLeuIleGlyValGlyValLeuSerCysValPheAsnSerLeuAlaGlyLysIleCysTyrAspAlaLeuAspProAlaLys   90

TyrAlaLysTrpLysProTrpLeuLysProTyrLeuAlaValCysIlePhePheAsnValIleLeuPheLeuValAlaLeuCysCysPhe   120

LeuLeuArgGlySerLeuGluSerThrLeuAlaTyrGlyLeuLysAsnGlyMetLysTyrTyrArgAspThrAspThrProGlyArgCys   150

PheMetLysLysThrIleAspMetLeuGlnIleGluPheLysCysCysGlyAsnAsnGlyPheArgAspTrpPheGluIleGlnTrpIle   180

SerAsnArgTyrLeuAspPheSerSerLysGluValLysAspArgIleLysSerAsnValAspGlyArgTyrLeuValAspGlyValPro   210

PheSerCysCysAsnProSerSerProArgProCysIleGlnTyrGlnLeuThrAsnAsnSerAlaHisTyrSerTyrAspHisGlnThr   240

GluGluLeuAsnLeuTrpLeuArgGlyCysArgAlaAlaLeuLeuAsnTyrTyrSerSerLeuMetAsnSerMetGlyValValThrLeu   270

LeuValTrpLeuPheGluValSerIleThrAlaGlyLeuArgTyrLeuHisThrAlaLeuGluSerValSerAsnProGluAspProGlu   300
                                   Peptide 4
CysGluSerGluGlyTrpLeuLeuGluLys [SerValProGluThrTrpLysAlaPheLeuGluSerPhe] LysLysLeuGlyLysSerAsn 330
          Peptide 5
GlnValGlu [AlaGluGlyAlaAspAlaGlyProAlaProGluAlaGly]                                           346
```

Figure 7 Amino acid sequence of *rds* protein. Residue number indicated to the right. Sequences used to direct the synthesis of peptides are indicated in boxes.

---

region of the *rds* cDNA was down-stream of the β-galactosidase gene in the same reading frame. When a bacterial culture that had been transformed with this plasmid was treated with isopropyl β-D-thiogalactoside (IPTG), the expression of a β-galactosidase-*rds* fusion protein was induced. An immunoblot was prepared from an extract of this transformed culture before and after IPTG-induction. The antipeptide-#4 sera was found to react with a band of the predicted molecular weight of the fusion protein in the induced but not the uninduced lane. This reactivity could also be abolished by pre-incubating the antipeptide-#4 sera with synthetic peptide #4.

Based upon these results, we concluded that antipeptide-#4 sera was detecting the *rds* protein in retinal extracts. By the same argument that was made for the mRNA, this protein is photoreceptor-specific in retina. To test for N-linked glycosylation of the *rds* protein, we pre-digested an aliquot of the two retinal extracts with endoglycosidase F (Elder and Alexander, 1982) prior to electrophoresis and transfer. The band detected by the antipeptide-#4 sera in the lane digested with endoglycosidase F was narrower and of lower molecular weight than the band detected in the undigested lane (data not shown). This result confirmed that the *rds* protein is glycosylated. Since it is not secreted (no signal peptide), it is likely that *rds* is an integral membrane protein. In addition, some degree of proteolytic processing of the *rds* protein is likely taking place, since the molecular weight of both glycosylated and unglycosylated forms is significantly less than the predicted molecular weight (39,259 D) of the unmodified protein. The antipeptide-#4 sera is presently being used for immunocytochemical localization of the *rds* protein within photoreceptors.

CONCLUSIONS

Cloning *rds* represents the first molecular description of a gene determining neuronal degeneration in mammals where nothing was known in advance about the gene product. Although technically gratifying, in a real sense the study of *rds* has scarcely begun. The promise offered by cloning the gene for a mutation is that a bridge can be spanned between the structure of the gene product and the functional disruption that results from the loss of that protein. With *rds*, we now know the structure of the gene product. We have begun to learn about its distribution within photoreceptors. From the work of others, we have learned a lot about the morphologic abnormalities seen in *rds* mutants. The bridge that we have yet to span is to learn the function of the *rds* protein, and to understand how loss of that function gives rise to the developmental and degenerative abnormalities seen in *rds/rds* photoreceptors.

Further effort in this direction is well worthwhile. Beyond the basic knowledge to be gained into the workings of photoreceptors, *rds* has significant clinical relevance. It bears important similarities to several forms of human retinitis pigmentosa. Understanding what causes the slow death of photoreceptors in *rds* may help us understand what causes the photoreceptor degeneration that leads to blindness in these human diseases. Thinking more broadly, neuronal degeneration is a process involved in several human inherited neurological disorders including familial Alzheimer's disease, Huntington's disease and the hereditary cerebellar ataxias. Since the photoreceptor is a type of neuron, it is possible that the biochemical events underlying gradual degeneration of neuronal populations in some of these other genetic diseases may be similar to the events leading to photoreceptor death in *rds/rds*.

REFERENCES

Birnstiel, ML, Busslinger, M and Strub, K (1985) Transcription termination and 3' processing: the end is in site! Cell **41**, 349-359

Carter-Dawson, LD, LaVail, MM and Sidman, RL (1978) Differential effect of the *rd* mutation on rods and cones in the mouse retina. Invest. Ophthalmol. Visual Sci. **17**, 489-498

Cohen AI (1983) Some cytological and initial biochemical observations on photoreceptors in retinas of *rds* mice. Invest Ophthalmol Vis Sci **24**, 832-843

Danielson, PE, Forss-Petter, S, Brow, MA, Calavetta, L, Douglas, J, Milner, RJ and Sutcliffe, JG (1988) p1B15: A cDNA clone of the rat mRNA encoding cyclophilin. DNA **7**, 261-267.

Demant P, Ivanyi D and Van Nie R (1979) The map position of the *rds* gene on the 17th chromosome of the mouse. Tissue Antigens **13**, 53-56

Elder, JH and Alexander, S (1982) *endo-β-N*-Acetylglucosaminidase F: Endoglycosidase from *Flavobacterium meningosepticum* that cleaves both high-mannose and complex glycoproteins. Proc. Natl. Acad. Sci. USA **79**, 4540-4544

Hawkins RK, Jansen HG and Sanyal S (1985) Development and degeneration of retina in *rds* mutant mice: photoreceptor abnormalities in the heterozygotes. Exp Eye Res **41**, 701-720

Hoggan, MD, Halden, NF, Buckler, CE and Kozak, CA (1988) Genetic mapping of the mouse *c-fms* proto-oncogene to chromosome 18. J. Virol. **62**, 1055-1056.

Jan, LY and Revel, JP (1974) Ultrastructural localization of rhodopsin in the vertebrate retina. J. Cell Biol. **62**, 257-273

Jansen HG and Sanyal S (1984) Development and degeneration of retina in *rds* mutant mice: electronmicroscopy. J comp Neurol **224**, 71-84

Kozak, M (1986) Point mutations define a sequence flanking the AUG initiator codon that modulates translation by eukaryotic ribosomes. Cell **44**, 283-292

Kozak, M (1984) Compilation and analysis of sequences upstream from the translational start site in eukaryotic mRNAs. Nucleic Acids Res. **12**, 857-872

LaVail, MM and Sidman, RL (1974) Retinal degeneration in the mouse. Arch. Ophthalmol. **91**, 394-400.

Mullen, RJ, Eicher, EM and Sidman, RJ (1976) Purkinje cell degeneration, a new neurological mutation in the mouse. Proc. Nat. Acad. Sci, USA **73**, 208-212

Papermaster, DS, Schneider, BG, Zorn, MA, Kraehenbuhl, JP (1978) Immunocytochemical localization of opsin in outer segments and Golgi zones of frog photoreceptor cells, J. Cell Biol. **77**, 196-210

Proudfoot, NJ and Brownlee, GG (1976) 3' non-coding region sequences in eukaryotic messenger RNA. Nature **263**, 211-214

Reuter JH and Sanyal S (1984) Development and degeneration of retina in *rds* mutant mice: The electroretinogram. Neurosci Lett **46**, 231

Ruther, U and Muller-Hill, B (1983) Easy identification of cDNA clones. The EMBO Journal **2**, 1791-1794

Sanyal S, Chader G and Aguirre G (1985) Expression of retinal degeneration slow (*rds*) gene in retinal of the mouse. In RETINAL DEGENERATION: EXPERIMENTAL AND CLINICAL STUDIES, pp 239-256 Alan R. Liss, Inc.

Sanyal, S, Dees, C and Zeilmaker, GH (1986) Development and degeneration of retina in *rds* mutant mice: observations in chimaeras of heterozygous mutant and normal genotype. J. Embryol. exp. Morph. **98**, 111-121

Sanyal S and Jansen H (1981) Absence of receptor outer segments in the retina of *rds* mutant mice. Neurosci Lett **21**, 23-26

Sanyal, S and Zeilmaker, GH (1984) Development and degeneration of retina in *rds* mutant mice: light and electron microscopic observations in experimental chimeras. Exp. Eye Res. **39**, 231-246

Sanyal S, DeRuiter A and Hawkins RK (1980) Development and degeneration of retina in *rds* mutant mice: light microscopy. J Comp Neurol **194**, 193-198

Sanyal, S, Fletcher, R, Liu, YP, Aguirre, G, and Chader, G (1984) Cyclic nucleotide contend and phosphodiesterase activity in *rds* mouse (020/A) retina. Exp. Eye Res. **38**, 247-256

Schalken JJ, Janssen JJ, de Grip WJ, Hawkins RK, Sanyal S (1985) Immunoassay of rod visual pigment (opsin) in the eyes of *rds* mutant mice lacking photoreceptor outer segments. Biochim Biophys Acta **839**, 122-126

Sidman RL and Green MC (1965) Retinal degeneration in the mouse: location of the *rd* locus in linkage group XVII. J Hered **56**, 23-29

Sidman, RL and Green, MC (1965) Retinal degeneration in the mouse: location of the *rd* locus in linkage group XVII. J. Hered. **56**, 23-39

Sidman, RL and Green, MC (1970) Nervous, a new mutant mouse with cerebellar disease. In Les mutants pathologiques chez l'animal, pp 69- 79, Sabourdy, ed. Centre National de la Recherche Scientific, Paris

Travis, GH and Sutcliffe, JG (1988) Phenol emulsion-enhanced DNA-driven subtractive cDNA cloning: isolation of low-abundance monkey cortex-specific mRNAs. Proc. Natl. Acad. Sci. USA **85**, 1696-1700.

Travis GH, Brennan MB, Danielson PE, Kozak CA and Sutcliffe JG (1989) Identification of a photoreceptor-specific mRNA encoded by the gene responsible for retinal degeneration slow (*rds*). Nature **338**, 70-73

Van Nie R, Ivanyi D, and Demant P (1978) A new H-2 linked mutation, *rds*, causing retinal degeneration in the mouse. Tissue Antigens **12**, 106-108

Van Gulik PJ and Kortweg R (1940) Susceptibility to follicular hormone and disposition to mammary cancer in female mice. Am J Cancer **38**, 506

Molecular Biology of the Retina: Basic and
Clinically Relevant Studies, pages 115–137

INTERPHOTORECEPTOR RETINOID-BINDING PROTEIN: BIOCHEMISTRY AND MOLECULAR BIOLOGY[1]

Gregory I. H. Liou, Li Geng, and Wolfgang Baehr

Department of Ophthalmology, Cullen Eye Institute, Baylor College of Medicine, Houston, TX 77030

INTRODUCTION

Vision is initiated upon the absorption of photoenergy by the chromophore of the visual pigment rhodopsin. The chromophore is the 11-cis isomer of retinal (vitamin A aldehyde). Absorption of light causes isomerization of 11-cis retinal to all-trans retinal, which is then reduced to all-trans retinol and released from the photoreceptor cells. The reactions transforming all-trans retinol to 11-cis retinal which are required to regenerate rhodopsin (reviewed by Bok, 1985) are known to occur in the pigment epithelium (PE) where a retinal isomerase has recently been identified (Bernstein et al., 1987; Deigner et al., 1989).

The PE is separated from the photoreceptor cells by a space filled with the interphotoreceptor matrix (IPM), which is an extracellular matrix composed of proteoglycans and glycoproteins and through which retinoid isomers have to be transported. Because of the membranolytic effect of retinol and its relative insolubility in aqueous media, the trans-

---

[1]This work was supported in part by the following awards: a grant from The National Eye Institute (EY03829) to G.I.L.; grants from the National Retinitis Pigmentosa Foundation (RP Foundation Fighting Blindness) to W.B. and G.I.L.; a Jules and Doris Stein Professorship from Research to Prevent Blindness, Inc. to W.B.; and a grant of the Advanced Technology Program of the Texas Higher Education Coordinating Board to G.I.L. and W.B.

port of retinol requires specific binding proteins. In the plasma, its transport is mediated by binding to serum retinol binding protein (SRBP), which is coupled as a 1:1 molar complex to transthyretin (prealbumin) (Goodman, 1984). Both transthyretin and SRBP, which are thought to be synthesized in the liver, have recently been shown to be also present in PE cells (Martone et al., 1988). In the neural retina, additional retinoid-binding proteins have been identified, including cytosolic retinol and retinoic acid binding proteins (CRBP and CRABP) (Saari et al., 1978), and binding proteins carrying both 11-cis retinol and retinal (CRALBP) (Stubbs et al., 1979; Crabb et al., 1988a; Crabb et al., 1988b). CRBP, but not CRABP, has also been found in PE cells (Saari et al., 1978). In the IPM, two retinoid binding proteins have been characterized: a high molecular weight protein, termed interphotoreceptor or interstitial retinoid-binding protein (IRBP), as a prominent constituent of the mammalian retina (Wiggert et al., 1977; Adler and Martin 1982; Lai et al., 1982; Liou et al., 1982) and a low molecular weight protein, purpurin, in the chick retina (Berman et al., 1987). IRBP appears to be unrelated to the other known retinoid binding proteins. Purpurin has 50% sequence homology to human SRBP and was suggested to serve as a neuro-retina adhesion and cell survival protein (Schubert et al., 1986).

IRBP (7S protein) was discovered more than 10 years ago in bovine retina (Wiggert et el., 1977). It binds retinol in a light-dependent manner, but also weakly binds other retinoids and fatty acids. Its precise function as a retinoid carrier (Adler and Martin, 1982; Liou et al., 1982) or as a passive reservoir for retinoids (retinoid buffer protein, Ho et al., 1989) remains to be determined. Transport and metabolism of retinoids have most recently been reviewed by Bridges et al. (1988), and specifically IRBP by Chader (1989). This chapter will briefly summarize our current knowledge concerning the biochemistry of IRBP and the nature of its ligands, and discuss the primary sequence and gene structure of bovine and human IRBP in more detail.

BIOSYNTHESIS AND EXTRACELLULAR LOCATION OF IRBP

IRBP was shown by immunocytochemical methods to be present in the extracellular interphotoreceptor matrix of various vertebrate retinas (Bunt-Milam and Saari, 1983).

Later it was demonstrated by in-situ hybridization that IRBP is synthesized in and secreted from photoreceptor cells (Van Veen et al., 1986). This is consistent with the finding that IRBP could not be detected in a patient with advanced Retinitis Pigmentosa whose retina was essentially devoid of photoreceptors (Bergsma et al, 1977), and in animals with degenerated photoreceptor cells such as rd mice (Carter-Dawson et al., 1986) and RCS rats (Gonzalez-Fernandez et al., 1984).

During normal development of the retina, IRBP is first detected at the time when the photoreceptor inner segments start to differentiate (Carter-Dawson et al., 1986; Johnson et al., 1985). Its secretion to the IPM appears to coincide with the onset of outer segment elongation (Van Veen et al., 1988) emphasizing its involvement in vision. IRBP is also expressed in the mammalian pineal gland (Bridges et al., 1986a) which is thought to be a rudimentary photoreceptor. Levels of expression in the pineal gland, however, are significantly reduced and IRBP is apparently not secreted (Rodrigues et al., 1986). Retinoblastoma tumor cells which derive from cone photoreceptors (Bogenmann et al., 1988) also express IRBP (Kyritsis et al., 1985; Fong et al., 1988a). Presently it is unknown whether the IRBP synthesized by the cone photoreceptors is identical to or distinct from the one synthesized by the rods.

BIOCHEMICAL ANALYSIS OF IRBP

IRBP is a single subunit lipo-glycoprotein. The molecular weight of mature bovine IRBP (Borst et al., 1989) as calculated from its predicted 1264 residue amino acid sequence, excluding retinol, carbohyrate and fatty acids, is 137,300. This value is roughly in agreement with its mobility of 145 kd on SDS gels (Fong et al., 1984a), and its apparent molecular weight of 133,000 determined by sedimentation equilibrium (Adler et al., 1985; Saari et al., 1985). The molecular weight of human IRBP (1230 residues) is 133,400 as calculated from its cDNA sequence (Liou et al., 1989). The lower molecular weight of human IRBP is largely due to a shortened C-terminus (Liou et al., 1989). The shape of the molecule appears to be elongated, as judged by the relation of its Stokes' radius to its molecular mass, its sedimentation properties, and by its appearance in electron-microscopical specimens (Adler et al., 1987). These observations provide a reasonable explanation for the overestimation of the molecular weight of native IRBP on gel-filtration (Liou et al., 1982; Adler and Martin, 1982). Partial removal of the carbohydrate chains of IRBP with peptide-N-glycosidases did not alter its unusual shape (Taniguchi et al., 1986; Adler et al., 1987).

Bovine IRBP has 152 acidic (glutamic and aspartic acid) and 100 basic (lysine and arginine) residues; human IRBP has only 143 acidic and 84 basic residues. The excess of acidic amino acids is consistent with the measured acidic pI of the proteins, which ranges from 4.4 to 4.8 (Fong et al., 1984a). At least four distinct isoelectric species have been described for human IRBP. The distribution of positive and negative charges appears to be uniform. The content of hydrophobic amino acids (aromatic, ile, leu, met, val) is 34-34.5%, which is substantially lower than observed in the intrinsic membrane protein opsin (43.4% in bovine opsin). IRBP has a low amount of cysteines (less than one percent, 9 cysteines in bovine, 7 cysteines in human IRBP), compared with 3% in bovine opsin.

Like opsin, IRBP is processed through the Golgi apparatus of photoreceptor cells. Extracellular bovine IRBP contains about 9 % carbohydrate (Fong et al., 1984a; Adler et al., 1985). The carbohydrate chains contain one to four sialic acid moieties presumably altering the apparent pI of the processed IRBPs and offering a possible explanation for the various isoelectric species of the mature, secreted proteins. No galactosamine was detected (Taniguchi et al., 1986), suggesting that the carbohydrate chains are asparagine (N)-linked which is consistent with the occurrence of several putative glycosylation sites NXT(S) found in the IRBP primary sequence (Borst et al., 1989). The carbohydrate chains are either the complex biantennary or a hybrid type. The role of the carbohydrate in the function and turnover of IRBP is unknown. We do know, however, that the carbohydrate chains are neither required for retinol binding (Taniguchi et al., 1986; Adler et al., 1987) nor for secretion, since retinas incubated with labelled carbohydrate in the presence of tunicamycin (which inhibits attachment of carbohydrate to asn) do secrete nonglycosylated IRBP into the medium (Fong et al., 1984b).

The quantity of IRBP in bovine IPM was determined by rocket immunoelectrophoresis using combined retina and PE rinses in agarose gels containing rabbit antibovine IRBP antiserum (Bridges et al., 1984). For an eye with a neural retina surface of 25 cm^2, the amount of IRBP was about 3 nmoles (Fong et al., 1984a) or 420 ug based on a molecular weight of 140,000. From 100 bovine eyes, roughly 14 mg IRBP (Fong et al., 1984a) may be isolated, approaching yields available for opsin (Papermaster and Dreyer, 1974).

LIGANDS BOUND TO IRBP

The endogenous ligands bound to IRBP are retinol and retinal isomers. As shown by several groups (Wiggert et al. 1979; Lai et al., 1982; Adler and Martin 1982; Liou et al. 1982; Adler and Evans, 1985; Saari et al., 1985), the amount of bound all-trans retinol, but not that of other ligands, increased when eyes were illuminated prior to isolation of IRBP (Fig. 1). In contrast, the amount of 11-cis retinal increased in IRBP isolated from dark-adapted eyes (Adler and Evans, 1985). One mole of purified bovine IRBP binds about 2 moles of exogenous all-trans retinol after illumination (Fong et al., 1984a; Saari et al., 1985). Thus, 3 nmoles of IRBP in the light-adapted eye can bind up to 6 nmoles of

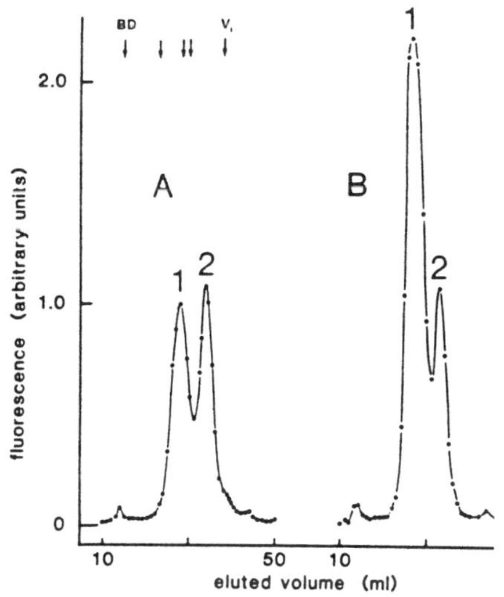

Figure 1. Comparison of the gel-filtration analysis of IPM from non-illuminated (A) and illuminated (B) bovine eyes. Fractions from a Sepharose 4B-CL column were monitored by their fluorescence emission at 470 nm (excitation at 330 nm). Peak 1, IRBP; Peak 2, CRBP. Arrows represent the elution positions of Blue Dextran (BD), thyroglobulin (669 kd), catalase (232 kd), aldolase (158 kd) and 3-nitrotyrosine (V_i). (Bridges et al., 1984, based on data of Liou et al., 1982). Reprinted with permission of *Vision Research*.

all-trans retinol, which is about 40% of the retinol released if all rhodopsin present in the retina were complete-ly bleached. When freshly isolated from light-exposed eyes, IRBP contains only up to 0.6 molecules of all-trans retinol, presumably due to weak ligand binding. The dissociation constant of bovine IRBP for all-trans retinol is approximately 10^{-6} M (Adler et al., 1985). This value is approximately two orders of magnitude weaker than observed for CRBP (Wiggert et al., 1977). This comparatively weak binding may be of advantage in a passive transport system in which retinol is passed on in the IPM from IRBP to IRBP ("bucket brigade", Chader, 1989).

Bovine IRBP can bind exogenously added retinoic acid, cholesterol, and tocopherol (Fong et al., 1984a), the physiological significance of which is unknown. In addition to retinoids, IRBP can bind fatty acids, e.g. monkey IRBP was found to contain 6.51 moles of fatty acid per mole of protein (Bazan et al., 1985). Of these fatty acids, 65% were found to be noncovalently attached, with the remainder covalently bound. These fatty acids account for 90% of the total fatty acid bound to IPM proteins extracted. Thus, IRBP may also function as an extracellular fatty acid carrier.

GENE STRUCTURE OF IRBP

Recently, partial cDNA sequences encoding bovine and human IRBP (Barret et al., 1985; Liou et al., 1986; Liou et al., 1987b; Fong and Bridges, 1988b) and genomic clones containing the complete IRBP structural gene (Borst et al., 1989; Liou et al., 1989) have been described. The human gene was assigned to chromosome 10 (Liou et al., 1987a). The human genomic IRBP clone HGL-3 (Liou et al, 1989) and the two bovine genomic clones lambda-gIRBP7, lambda-gIRBP8 (Borst et al., 1989) contained the entire IRBP gene, including several kb upstream and at least several hundred bp downstream non-coding sequences. The structural genes of bovine and human IRBP are compact, spanning less than 12 kb, and consisting of 4 exons interrupted by 3 introns. The positions of the introns in both genes are identical. In the following paragraphs, characteristic data of both genes are highlighted and compared (human IRBP gene data are taken from Liou et al., 1989, bovine data are from Borst et al., 1989).

The transcription start site (5′ end of the gene) was located in human at an adenine, 122 bp upstream of the translation initiation codon ATG, and in bovine at a cytosine, 125 bp upstream. No suitable "TATA" or "CCAAT"-box consensus sequences (Breathnach and Chambon, 1981) could be found within conventional distances upstream from the transcription start site of either gene. To identify functional 5′-upstream regulatory sequences that are able to drive photoreceptor-specific expression of IRBP or foreign antigens, transfections of various constructs containing upstream sequences of human IRBP are currently being performed in one of the author's laboratory (G.I.L.) using retinoblastoma cell lines and transgenic mice. Preliminary results indicate that regulatory elements conferring tissue-speci-

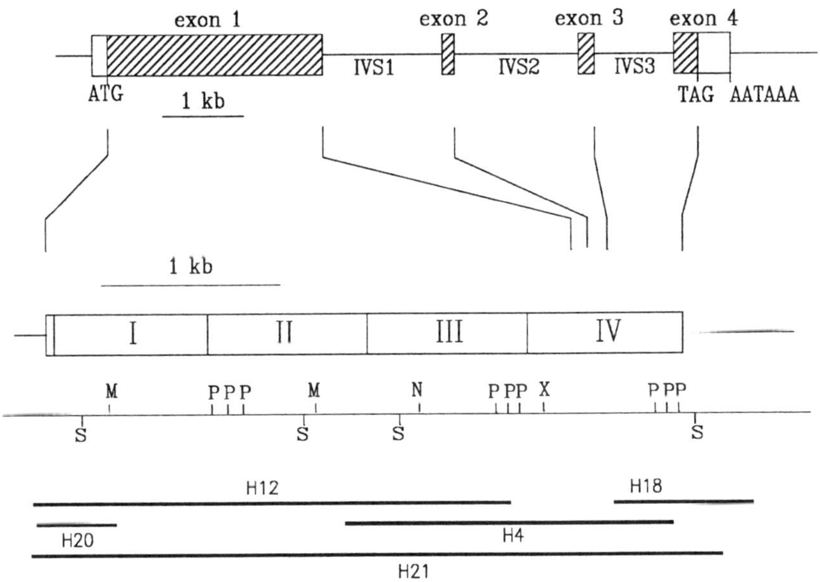

Figure 2. Map of human IRBP gene and its mRNA. The quadruplicated mRNA segments are depicted as I-IV, and the extent of four overlapping cDNA fragments H20, H12, H4, H18, and a full-length clone H21 is indicated by bars. The box 5′ to the first quadruplicated segment represents sequences translated into a signal peptide. IVS, intervening sequences. Restriction sites of IRBP mRNA: P, PstI; M, SmaI; S, SstI; X, XhoI; N, NcoI.

ficity of IRBP expression reside in an upstream sequence spanning less than 2 kb.

In both the human and bovine IRBP genes, the three introns (1.3 to 2.5 kb in length) are located near the 3'-end of the coding sequence. The large first exon contains more than three quarters of the structural gene. It extends from the transcription start site to the first intron junction consisting of 3176 bp in human (3173 in bovine). Thus, calculated from ATG, the coding sequence of bovine exon 1 is 6 bp shorter than that of human exon 1, or two amino acid residues are deleted in the predicted bovine peptide (see Fig. 3). The two central exons are short (192, and 143 bp, respectively) and identical in length in human and bovine IRBP. Human exon 4 is 750 bp, and bovine exon 4 is 2447 bp. This difference is due to a 1700 bp insertion in the 3'-untranslated region of the bovine gene (Si et al., 1989) and provides an explanation for the difference in lengths of the polyadenylated mRNAs of bovine and human IRBP. The major IRBP mRNAs were found on Northern blots to be 4.4 kb in human and 6kb in bovine (see Fig. 5), in agreement with the distances between the respective transcription start sites and the polyadenylation signals. Computer alignment of human and bovine cDNAs from transcription start to the translation stop codon reveals 84% homology.

INTERNAL GENE QUADRUPLICATION

The human IRBP cDNA sequence was compared to DNA and protein sequences contained in GenBank (release 56) and PIR (release 17) data banks. The comparison yielded no gene or cDNA sequences with significant similarity to IRBP. Dot matrix analysis of bovine and human IRBP mRNA with itself, however, strongly suggests that the gene arose by quadruplication (Fong and Bridges, 1988b; Liou et al., 1989) of an ancestral sequence of approximately 900 bp. Rat and human SRBPs also have been suggested to be the product of at least one gene duplication. The four gene segments in IRBP reveal a high degree of similarity preserving several domains of the expressed protein (see below). In the IRBP gene, all intron junctions were found in the last quadruplicated segment. This arrangement of introns generates one large exon containing the first three segments of IRBP and parts of the fourth, one exon containing the C-terminal sequences, and two small exons exclusively containing sequences of the fourth segment. In rat SRBP, whose tertiary structure has been determined by X-ray crystallography (Laurent et al., 1985), exons correspond to discrete tertiary structural elements. In rhodopsin, intron positions roughly coincide with membrane boundaries (Nathans and Hogness, 1984). For IRBP, it is difficult to comprehend that this arrangement of exon/intron junctions reflects the borders of building blocks. A final answer will have to await more detailed information about its tertiary structure.

PRIMARY SEQUENCE OF IRBP

The N-terminus of human and rhesus monkey IRBP was determined by Edman degradation of purified IRBP as GPTHLFQPSL, that of bovine IRBP as FQPSL (Fong et al., 1986) and is thus five residues shorter (Fig. 3). This loss of five residues may be due to differences in posttranslational processing (Borst and Nickerson, 1988). In human, the 17-amino acid signal peptide preceding the mature N-terminus glycine is rich in hydrophobic amino acids and ends in an alanine which is most commonly found at this position (von Heijne, 1986). The alanine is replaced by threonine in bovine providing a possible explanation for the differences in N-terminal processing. The predicted human IRBP C-terminus is 41 amino acids shorter than that of bovine IRBP (Liou et al., 1989). This accounts for part of the difference in molecular weight between the two proteins. 81.4% identitities were observed when the complete mature human sequence was compared with the bovine IRBP sequence (Fig. 3).

As indicated by the quadruplication of the gene sequence, analysis of the primary sequence of IRBP reveals that this protein consists of four homologous segments of approximately 300 residues each. An optimized alignment of the four segments of human IRBP (36% identities) is shown in Fig. 3. Two large domains, spanning 71 and 85 residues, respectively, are the most conserved domains in the four segments, uninterrupted by gaps, deletions or insertions (Fig. 3). The first domain contains two invariant sequences (o-gly-tyr-leu-arg-o-asp, and o-o-o-asp-o-arg, where o = hydrophobic residue, indicated by O-G-Y-L-R-O-D and O-O-O-D-O-R above the sequence in Fig. 3) both of which have an excess of hydrophobic residues. On the basis of their exact conservation in the four segments of IRBP, both in human and bovine, we speculate that these domains may be involved in retinoid binding.

```
      +1
I      G P T H L F Q P S L V L D M A K V L L D N Y C F P   25
II       T L R S A L P G V V H C L Q E V L K D Y Y T L V   327
III      E F H Q S L G A L V E G T G H L L E A H Y A R P   637
IV       A L R A K V P T V L Q T A G K L V A D N Y A S A   938

I      E N L L G M Q E A I Q Q A I K - - S H E I L S I S   48
II     D R V P - - - - - T L L Q H L A - - S M D F S T V V 346
III    E V V G Q T S A L L R A K L A - - Q G A Y R T A V   660
IV     E L G - - - - - A K M A T K L S G L Q S R Y S R V T 959

I      D P Q T L A S V L T A G V Q S S L N D P R L V I -   72
II     S E E D L V T E L N A G L Q A A S E D P R L L V R   371
III    D L E S L A S Q L T A D L Q E V S G D H R L L V -   684
IV     S E V A L A E I L G A D L Q M L S G D P H L K A -   983

I        - - - - - - - - - - S Y E P S T P E P P P Q V P   86
II       A I G P T E T P S W P A P D A A A E D S P G V A   395
III      - - - - F H S P G E L V V E E A P P P P P - A V   703
IV       - - - - - A H I P E N A K D R I P G I V P M Q I   1002

                               ├domain 1→
I      A L T S L S E E E L L A W L Q R G L R H E V L E G   111
II     P E L P E D E A I R Q A L V D S V F Q V S V L P G   420
III    P S - P E E L T Y - - - L I E A L F K T E V L P G   724
IV     P S - P E V F E E - - - L I K F S F H T N V L E D   1023

         O-G-Y-L-R-O-D
I      N V G Y L R V D S V P G Q E V L S M M G E F L V A   136
II     N V G Y L R F D S F A D A S V L G V L A P Y V L R   445
III    Q L G Y L R F D A M A E L E T V K A V G P Q L V R   749
IV     N I G Y L R F D M F G D G E L L T Q V S R L L V E   1048

                             O-O-O-D-O-R
I      H V W G N L M G T S A L V L D L R H C T G G Q V S   161
II     Q V W E P L Q D T E H L I M D L R H N P G G P S S   470
III    L V W Q Q L V D T A A L V I D L R Y N P G S Y S T   774
IV     H I W K K I M H T D A M I I D M R F N I G G P T S   1073

           ←domain 1┤                 ├domain 2→
I      G I P Y I I S Y L H - P G N T I L H V D T I Y N R   185
II     A V P L L L S Y F Q G P E A G P V H L F T T Y D R   495
III    A I P L L C S Y F F - E A E P R Q H L Y S V F D R   798
IV     S I P I L C S Y F F D E G P - P V L L D K I Y S R   1097

I      P S N T T T E I W Q L P Q V L G E R Y G A D K D V   210
II     R T N I T Q E H F S H M E L P G P R Y S T Q R G V   520
III    A T S K V T E V W T L P Q V A G Q R Y G S H K D L   823
IV     P D D S V S E L W T H A Q V V G E R Y G S K K S M   1122
```

```
I      V V L T S S Q T R G V A E D I A H I L K Q M R R A   235
II     Y L L T S H R T A T A A E E F A F L M Q S L G W A   545
III    Y I L M S H T S G S A A E A F A H T M Q D L Q R A   848
IV     V I L T S S V T A G T A E E F T Y I M K R L G R A   1147

                                     ◄—domain 2 —|
I      I V V G E R T G G G A L D L R K L R I G E S D F F   260
II     T L V G E I T A G N L L H T R T V P L L D T P E G   570
III    T V I G E P T A G G A L S V G I Y Q V A S S P L Y   873
IV     L V I G E V T S G G C Q P P Q T Y H V D D T N L Y   1172

I      F - - - T V P V S R S L G P L G G G S Q T W E G S   282
II     S L A L T V P V - - - L T F I D N H G E A W L G G   592
III    A - - - S M P T Q M A M S A T T - - G K A W D L A   893
IV     - - - L T I P T A R S V G A S D - - G S S W E G V   1192

I      G V L P C V G T P A E Q A L E K A L A I L           303
II     G V V P D A I V L A E E A L D K A Q E V L           613
III    G V E P D I T V P M S E A L S I A Q D I V           914
IV     G V T P H V V V P A E E A L A R A K E M L           1213

C-TERMINUS  Q H N Q L R V K R S P G L Q D H L              1230
```

Figure 3. Comparison of the four repeated segments of human IRBP. Only residues conserved in all segments are boxed. The following residues are considered identical ("conservative substitutions"): I=L=V=M; K=R; D=E (Dayhoff et al., 1983). Asn-linked glycosylation sites and proline-rich sequences are underlined. The beginning and end of two large domains which are not interrupted by gaps and contain most of the conserved residues are indicated by arrows.

Interspecies comparison of individual segments shows that segment III is the most invariant, with only 14 differing residues in human and bovine IRBP. In contrast, segments I, II and IV have 42, 47, and 32 substitutions, respectively. The most variable domain (apart from the differing C-termini) is located in segment II (position 368 to 395 in the human sequence). Of 27 residues, only 11 can be aligned with the corresponding bovine IRBP sequence. Moreover, the bovine sequence is shortened in this region by two residues, the only internal deletion in IRBP.

segment I:

```
h GPTHLFQPSLVLDMAKVLLDNYCFPENLLGMQEAIQQAIKSHEILSISDPQT  52
b      FQPSLVLEMAQVLLDNYCFPENLMGMQGAIEQAIKSQEILSISDPQT  47
  *****          *                *  *     *

h LASVLTAGVQSSLNDPRLVISYEPSTPEPPPQVPALTSLSEEELLAWLQRGL 104
b LAHVLTAGVQSSLNDPRLVISYEPSTLEAPPRAPAVTNLTLEEIIAGLQDGL  99
    *                       * *  **    *  *     *  *

h RHEVLEGNVGYLRVDSVPGQEVLSMMGEFLVAHVWGNLMGTSALVLDLRHCT 156
b RHEILEGNVGYLRVDDIPGQEVMSKLRSFLVANVWRKLVNTSALVLDLRHCT 151
                 *        * **    *  **  *

h GGQVSGIPYIISYLHPGNTILHVDTIYNRPSNTTTEIWTLPQVLGERYGADK 208
b GGHVSGIPYVISYLHPGSTVSHVDTVYDRPSNTTTEIWTLPEALGEKYSADK 203
    *              *  *      *             **     *

h DVVVLTSSQTRGVAEDIAHILKQMRRAIVVGERTGGGALDLRKLRIGESDFF 260
b DVVVLTSSRTGGVAEDIAYILKQMRRAIVVGERTVGGALNLQKLRVGQSDFF 255
          * *       *               *    * *     *

h FTVPVSRSLGPLGGGSQTWEGSGVLPCVGTPAEQALEKALAIL          303
b LTVPVSRSLGPLGEGSQTWEGSGVLPCVGTPAEQALEKALAVL          298
  *            *
```

segment II:

```
h TLRSALPGVVHCLQEVLKDYYTLVDRVPTLLQHLASMDFSTVVSEEDLVTKL 355
b MLRRALPGVIQRLQEALREYYTLVDRVPALLSHLAAMDLSSVVSEDDLVTKL 350
  *  *      **   *            *  *   *  *

h NAGLQAASEDPRLLVRAIGPTETPSWPAPDAAAEDSPGVAPELPEDEAIRQA 407
b NAGLQAVSEDPRLQVQVVRPKEASSGPEEEA--EEPPEAVPEVPEDEAVRRA 400
        *      * ** * * ** * **  **  * ***          *

h LVDSVFQVSVLPGNVGYLRFDSFADASVLGVLAPYVLRQVWEPLQDTEHLIM 459
b LVDSVFQVSVLPGNVGYLRFDSFADASVLEVLGPYILHQVWEPLQDTEHLIM 452
                               *  *    *

h DLRHNPGGPSSAVPLLLSYFQGPEAGPVHLFTTYDRRTNITQEHFSHMELPG 511
b DLRQNPGGPSSAVPLLLSYFQSPDASPVRLFSTYDRRTNITREHFSQTELLG 504
     *                 *   *  *            *    **  *

h PRYSTQRGVYLLTSHRTATAAEEFAFLMQSLGWATLVGEITAGNLLHTRTVP 563
b RPYGTQRGVYLLTSHRTATAAEELAFLMQSLGWATLVGEITAGSLLHTHTVS 556
  ** *                   *                   *    *  *

h LLDTPEGSLALTVPVLTFIDNHGEAWLGGGVVPDAIVLAEEALDKAQEVL   613
b LLETPEGGLALTVPVLTFIDNHGECWLGGGVVPDAIVLAEEALDRAQEVL   606
         *                *
```

segment III:

```
h EFHQSLGALVEGTGHLLEAHYARPEVVGQTSALLRAKLAQGAYRTAVDLESL 665
b EFHRSLGELVEGTGRLLEAHYARPEVVGQMGALLRAKLAQGAYRTAVDLESL 659
     *   *                     **

h ASQLTADLQEVSGDHRLLVFHSPGELVVEEAPPPPPAVPSPEELTYLIEALF 717
b ASQLTADLQEMSGDHRLLVFHSPGEMVAEEAPPPPPVVPSPEELSYLIEALF 710
                             *        *

h KTEVLPGQLGYLRFDAMAELETVKAVGPQLVRLVWQQLVDTAALVIDLRYNP 769
b KTEVLPGQLGYLRFDAMAELETVKAVGPQLVQLVWQKLVDTAALVVDLRYNP 762
                                 *

h GSYSTAIPLLCSYFFEAEPRQHLYSVFDRATSKVTEVWTLPQVAGQRYGSHK 821
b GSYSTAVPLLCSYFFEAEPRRHLYSVFDRATSRVTEVWTLPHVTGQRYGSHK 814
                      *                    * *

h DLYILMSHTSGSAAEAFAHTMQDLQRATVIGEPTAGGALSVGIYQVGSSPLY 873
b DLYVLVSHTSGSAAEAFAHTMQDLQRATIIGEPTAGGALSVGIYQVGSSALY 866
                                                   *

h ASMPTQMAMSATTGKAWDLAGVEPDITVPMSEALSIAQDIV            914
b ASMPTQMAMSASTGEAWDLAGVEPDITVPMSVALSTARDIV            907
                *                *     *
```

segment IV:

```
h ALRAKVPTVLQTAGKLVADNYASAELGAKMATKLSGLQSRYSRVTSEVALAE 966
b TLRAKVPTVLQTAGKLVADNYASPELGVKMAAELSGLQSRYARVTSEAALAE 959
  *                      *   *   **              *

h ILGADLQMLSGDPHLKAAHIPENAKDRIPGIVPMQIPSPEVFEELIKFSFHT 1018
b LLQADLQVLSCDPHLKTAHIPEDAKDRIPGIVPMQIPSPEVFEDLIKFSFHT 1011
    *             *     *

h NVLEDNIGYLRFDMFGDGELLTQVSRLLVEHIWKKIMHTDAMIIDMRFNIGG 1070
b NVLEGNVGYLRFDMFGDCELLTQVSELLVEHVWKKIVHTDALIVDMRFNIGG 1063
      *            *       *

h PTSSIPILCSYFFDEGPPVLLDKIYSRPDDSVSELWTHAQVVGERYGSKKSM 1122
b PTSSISALCSYFFDEGPPILLDKIYNRPNNSVSELWTLSQLEGERYGSKKSM 1115
       *                   *  **       **  *

h VILTSSVTAGTAEEFTYIMKRLGRALVIGEVTSGGCQPPQTYHVDDTNLYLT 1174
b VILTSTLTAGAAEEFTYIMKRLGRALVIGEVTSGGCQPPQTYHVDDTDLYLT 1167
            *                                    *

h IPTARSVGASDGSSWEGVGVTPHVVVPAEEALARAKEML              1213
b IPTARSVGAADGSSWEGVGVVPDVAVPAEAALTRAQEML              1206
           *          * * *    *  *  *
```

(continued)

C-terminus

```
h QHNQLRVKRSPGLQDHL----------------------------------- 1230
b QHTPLRARRSPRLHGRRKGHHRQSQGRAGSLGRNQGVVRPEVLTEAPSGQKR 1258
     **  *    * ***************************************

h ------
b GLLQCG
  ******
```

Figure 4. Comparison of the amino acid sequences of segments I-IV of human (h) and bovine (b) IRBP. Deviating residues are identified by an asterisk (I=L=V=M; K=R; D=E as in Fig. 3). Putative glycosylation sites and the five adjacent prolines are underlined. Conserved cysteine residues are marked by an arrow.

Of several putative glycosylation sites in the bovine IRBP sequence, only two are conserved when compared with the human sequence. These two sites are located in well-conserved domains in the central parts of segments I and II (Fig. 3, 4). Each of the four segments has a region rich in proline residues (Liou et al., 1989). In segment III, near the center of the IRBP molecule, a sequence of five adjacent prolines has been conserved in human and bovine IRBP (underlined in Fig. 3, 4). This positioning of a proline cluster near the center of the elongated molecule may be responsible for the formation of a "hinge" structure of IRBP as observed under the electron microscope (Adler et al., 1987). Another cluster of prolines in segment I of the human IRBP sequence is not conserved in bovine. Analysis of the cysteine distribution along the IRBP molecule reveals that six of seven cysteines of the human sequence are found at identical positions in the bovine sequence, emphasizing their importance for structure and function of IRBP. Hydrophobicity plots of the human IRBP sequence (Kyte and Doolittle, 1982) or hydrophobic moment analysis (Eisenberg, 1984) do not identify internal trans-membrane domains (results not shown). Secondary structure prediction (Chou and Fassman, 1978) indicates predominant beta-sheet character of the molecule. As observed in rat SRBP (Laurent et al., 1985), the retinol binding site in IRBP may be formed by a beta-barrel core consisting of several antiparallel beta-strands.

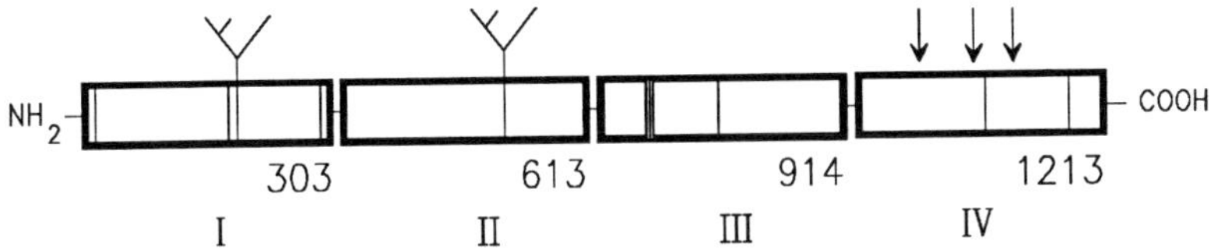

Figure 5. Simplified diagram of the primary structure of IRBP. The numbers denote the last amino acid of each segment I to IV. Approximate positions of conserved putative N-linked glycosylation sites, NXT(S), in segments I and II are marked with a branched structure. A proline-rich region in segment III, near the center of the molecule, is shaded. Vertical lines indicate conserved cysteines. Arrows indicate the position of introns.

There is no significant sequence similarity of IRBP with other retinol binding proteins. We noticed, however, two short segments with sequence similarity in rhodopsin and IRBP (Liou et al., 1989). First, the intradiscal (extracellular) loop connecting membrane spanning helices 2 and 3 of rhodopsin and the C-terminal end of IRBP's segment I share 15 matching residues in a 41 residue long segment. The sequence of this loop in rhodopsin has been precisely conserved in all mammalian opsins, but its function is unknown. Second, 8 matches in a 14 residue segment of the cytoplasmic loop 5/6 of opsin and the C-terminal segment of segment III have been found. Loop 5/6 of rhodopsin has been implicated in ROS G-protein binding, and serine-240 in opsin is known to be phophorylated by opsin kinase (for review, Baehr and Applebury, 1986). The significance of these similarities is not known.

DISTRIBUTION OF IRBP IN THE VERTEBRATES

IRBP appears to be widely distributed throughout the vertebrates (Wiggert et al., 1986; Bridges et al., 1986b). Immunoblots of IPM preparations from 20 species, probed with polyclonal antibodies raised against bovine IRBP, showed an immunoreactive, IRBP-like protein band (Bridges et al., 1986b). In the Osteichthyes (gizzard shad, black drum, southern flounder, sheepshead, red drum, and gold fish), the apparent molecular weight of the immunoreactive product was about 68 kd. In the other species, including Amphibia (Leopard frog, Bullfrog, toad, and salamander) ; Reptilia (turtle); Aves (chicken); Mammalia (cattle, human, rhesus monkey, squirrel monkey, rabbit, guinea pig, rat, mouse, and raccoon); and Chondrichthyes (stingray), the apparent molecular weight was around 134 kd. Among all the species, frog IRBP was very similar to mammalian protein in terms of its immunohistochemical localization, its molecular weight , retinol and concanavalin-A binding ability, and its being synthesized and secreted *in vitro* by the isolated retina. Goldfish IRBP bound all-trans retinol but did not bind concanavalin-A. The occurrence of IRBP-like proteins cross-reacting with bovine IRBP antibodies in the IPM of all 20 vertebrate species is consistent with the hypothesis that IRBP has evolved from a common ancestor as an important element in the vertebrate visual cycle.

Extensive sequence conservation among human, bovine, and mouse IRBPs (G.I.L., unpublished results) indicates that IRBP cDNA clones may be suitable probes for Northern and genomic Southern blotting. We have recently explored the potential of using human IRBP cDNAs as Northern hybridization probes to identify the distribution of IRBP-like mRNAs in various vertebrates. Using this procedure, we have demonstrated (Fig. 5) that the IRBP gene is transcribed in the retinas of human, bovine, mouse, rat, and frog. As shown in Fig. 5, the sizes of the mature, polyadenylated transcripts vary among species: 4.4 kb in human, 6.5 kb in bovine, 5.7 kb in mouse, and 4.4 kb in frog. The sizes of other mammalian IRBP mRNAs are 4.9 kb in guinea pig, 5.4-6.6 kb in rat, 6.3 kb in rabbit, and 7.6 kb in hamster (Inouye et al., 1989). Sequence comparison of the human and bovine IRBP genes has demonstrated that the difference in size of their single transcripts is due to a variation in the 3'-untranslated region. A similar variation will most likely explain the differences in sizes of the other mammalian transcripts, since the IRBP structural genes are predicted to be of the same size (IRBP peptides have nearly identical N-termini (Fong et al., 1986), and are roughly identical in size), and the 5'-untranslated areas in eucaryotes are generally short (0.1 kb) (Kozak, 1984). We were unable to identify any IRBP mRNA in 14 day embryonic or 9 weeks old chick retina. The same chick material, however, contains intact purpurin mRNA indicating the presence of undegraded mRNA. Northern hybridizations under relaxed stringency also failed to identify any IRBP mRNA in osteichtyes (data not shown).

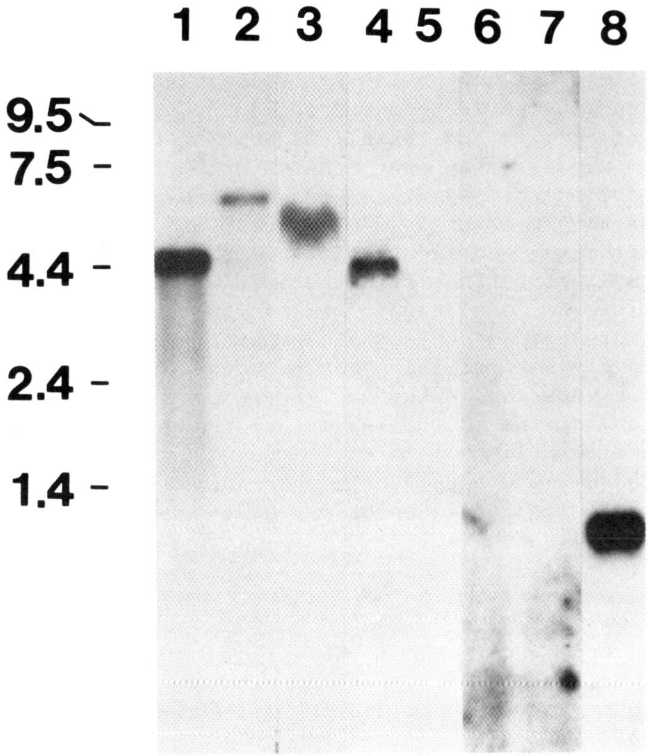

Figure 6. Identification of vertebrate IRBP mRNA by Northern blotting. Lane 1, human retina poly(A) mRNA ; lane 2, bovine retina poly(A) mRNA; lane 3, mouse retina total RNA; lane 4, frog retina poly(A) mRNA; lane 5, frog liver RNA (control); lane 6 and 8, chick retina poly(A) mRNA; lane 7, chick liver RNA. Hybridization probe for lanes 1-7 was radiolabelled full-length human IRBP cDNA H21 (Fig. 2); the probe for lane 8 was chick purpurin cDNA (a gift of Dr. David Schubert, The Salk Institute). High stringency hybridization conditions (5X SSC, 50% formamide at 42°C, washing at 65°C in 0.5X SSC) were used for lanes 1-3 and 8, and relaxed stringencies (5X SSC in 50% formamide at 37°C, washing at 50°C in 1X SSC) were used for lanes 4-7.

SUMMARY

Interphotoreceptor retinoid-binding protein (IRBP) is a recent and still somewhat enigmatic newcomer in the family of photoreceptor-specific proteins involved in vision. It has been isolated, characterized, cloned and sequenced from several species in less than a decade. Its extracellular localization and the increased amount of bound all-trans retinol following illumination are consistent with the proposed function of IRBP in extracellular transport or buffering of retinol and retinal isomers in the IPM. Its capacity to bind a great many other hydrophobic ligands, however, may indicate a multi-purpose function. Many questions concerning biosynthesis and processing, secretion, tertiary structure and ligand binding remain to be answered. Perhaps most interesting, from a molecular biology point of view, will be the regulation of transcription and tissue-specific expression of this photoreceptor protein, and possible involvement of its gene in human retinal degenerations. Constructs consisting of IRBP regulatory sequences and foreign antigens incorporated into the germ line of transgenic mice will help in the characterization of promoters and enhancer elements, while suppression of IRBP gene expression and its consequences may give us a more precise answer about IRBP function.

REFERENCES

Adler AJ, Martin KJ (1982). Retinol-binding proteins in bovine interphotoreceptor matrix. Biochem Biophys Res Comm 108:1601-1608.

Adler AJ, Evans, CD (1985). Proteins of the bovine interphotoreceptor matrix: retinoid binding and other functions. In Bridges CDB and Adler AJ (eds): "Progress in Clinical and Biological Research, Vol. 190. Interphotoreceptor Matrix in Health and Disease", New York: Alan R. Liss, pp 65-88.

Adler AJ, Evans CD, Stafford WF, III (1985). Molecular properties of bovine interphotoreceptor retinol-binding protein. J Biol Chem 260:4850-4855.

Adler AJ, Stafford WF, III, Slayter HS (1987). Size and shape of bovine interphotoreceptor retinoid-binding protein by electron microscopy and hydrodynamic analysis. J Biol Chem 262:13198-13203.

Baehr W, Applebury ML (1986). Exploring visual transduction

with recombinant DNA techniques. Trends in Neuroscience 9:198-203.

Barret DJ, Redmond TM, Wiggert B, Oprian DD, Chader GJ, Nickerson, JM (1985). cDNA clones encoding bovine interphotoreceptor retinoid-binding protein. Biochem Biophys Res Commun 131:1086-1093.

Bazan NG, Reddy TS, Redmond TM, Wiggert B, Chader GJ (1985). Endogenous fatty acids are covalently and noncovalently bound to interphotoreceptor retinoid-binding protein in the monkey retina. J Biol Chem 260:13677-13680.

Bergsma DR, Wiggert B, Funahashi M, Kuwabara T, Chader, GJ (1977). Vitamin A receptors in normal and dystrophic human retina. Nature 265:66-68.

Berman P, Gray P, Chen E, Keyser K, Ehrlich D, Karten H, LaCorbiere M, Schubert D (1987). Sequence analysis, cellular localization and expression of an neuroretina adhesion and cell survival molecule. Cell 51:135-142.

Bernstein PS, Law WC, Rando RR (1987). Isomerization of all-trans retinoids to 11-cis retinoids in vitro. Proc Nat Acad Sci USA 84:1849-1853.

Bogenmann E, Lochrie MA, Simon MI (1988). Cone cell-specific genes expressed in retinoblastoma. Science 240:76-78.

Bok D (1985). Retinal photoreceptor-pigment epithelium interactions. Invest Ophthalmol Vis Sci 26:1659-1694.

Borst DE, Nickerson JM (1988). The isolation of a gene encoding interphotoreceptor retinoid-binding protein. Exp Eye Res 47:825-838.

Borst DE, Redmond TM, Elser JE, Gonda MA, Wiggers B, Chader GJ, Nickerson JM (1989). Interphotoreceptor retinoid-binding protein: gene characterization, protein repeat structure, and its evolution. J Biol Chem 264:1115-1123.

Breathnach R, Chambon P (1981). Organization and expression of eucaryotic split genes coding for proteins. Ann Rev Biochem 50:349-383.

Bridges CDB, Alvarez RA, Fong SL, Gonzalez-Fernandez F, Lam DMK, and Liou GI (1984). Visual cycle in the mammalian eye: retinoid-binding proteins and the distribution of 11-cis retinoids. Vision Res 24:1581-1594.

Bridges CDB, Landers RA, Liou GI, Fong, SL (1986a). Interstitial retinol-binding protein (IRBP) in rat and bovine pineal organs: evolutionary vestige or functional molecule? In O'Brien PJ, Nein DC (eds): "Pineal and Retinal Relationships" New York: Academic Press, pp 383-399.

Bridges CDB, Liou GI, Alvarez RA, Landers RA, Landry, AM, Jr, Fong, SL (1986b). Distribution of interstitial retinol-bindingprotein (IRBP) in the vertebrates. J Exp Zool

239:335-346.

Bridges, CDB, Liou GI, Gonzales-Fernandez F, Alvarez R, Fong SL (1988). Vitamin A: Utilization, Metabolism, and Role in Retinal Disease. In Tso M (ed): "Retinal Diseases: Biomedical Foundations and Clinical Management. Philadelphia: JB Lippincott Company, pp 49-59.

Bunt-Milam AH, Saari, JC (1983). Immunocytochemical localization of two retinoid-binding proteins in vertebrate retina. J Cell Biol 97:703-712.

Carter-Dawson L, Alvarez RA, Fong SL, Liou GI, Sperling HG, Bridges CDB (1986). Rhodopsin, 11-cis vitamin A, and interstitial retinol-binding protein (IRBP) during retinal development in normal and rd mutant mice. Dev Biol 116:431-438.

Chader GJ (1989). Interphotoreceptor retinoid-binding protein (IRBP): A model protein for molecular biological and clinically relevant studies (Friedenwald lecture). Invest Ophthalmol Vis Sci 30:7-22.

Chou PY, Fasman GD (1978). Empirical predictions of protein conformation. Ann Rev Biochem 47:251-276.

Crabb JW, Johnson CM, Carr SA, Armes LG, Saari JC (1988a). The Complete Primary Structure of the Cellular Retinaldehyde-binding Protein from Bovine Retina. J Biol Chem 263:18678-18687.

Crabb JW, Goldflam S, Harris SE, Saari JC (1988b). Cloning of the cDNAs encoding the cellular retinaldehyde-binding protein from bovine and human retina and comparison of the protein structures. J Biol Chem 263:18688-18692.

Dayhoff MO, Barker WC, Hunt LT (1983). Establishing homologies in protein sequences. Methods Enzymol 91:534-545.

Deigner PS, Law WC, Canada FJ, Rando RR (1989). Membranes as the energy source in the endergonic transformation of vitamin A to 11-cis-retinol. Science 244:968-974.

Eisenberg D (1984). Three-dimensional structure of membrane and surface proteins. Ann Rev Biochem 53:595-623.

Fong SL, Liou GI, Landers RA, Alvarez RA, Bridges CDB (1984a). Purification and characterization of a retinol-binding glycoprotein synthesized and secreted by bovine neural retina. J Biol Chem 259:6534-6542.

Fong SL, Liou GI, Landers RA, Alvarez RA, Gonzalez-Fernandez F, Hazebrook PA, Lam DM-K, Bridges CDB (1984b). The characterization, localization, and biosynthesis of an interstitial retinol-binding glycoprotein in the human eye. J Neurochem 42:1667-1676.

Fong S-L, Cook RG, Alvarez RA, Liou GI, Landers RA, Bridges CDB (1986). N-terminal sequence homologies in interstitial retinoid-binding proteins from 10 vertebrate species. FEBS Letters 205:309-312.

Fong SL, Balakier H, Canton M, Bridges CDB, Gallie B (1988a). Retinoid-binding proteins in retinoblastoma tumors. Cancer Research 48:1124-1128.

Fong SL, Bridges, CDB (1988b). Internal quadruplication in the structure of human interstitial retinol-binding protein deduced from its cloned cDNA. J Biol Chem 263:14330-15334.

Gonzalez-Fernandez F, Landers RA, Glazebrook PA, Fong SL, Liou GI, Lam DMK, Bridges CDB (1984). An extracellular retinol-binding glycoprotein in the eyes of mutant rats with retinal dystrophy: Development, localization and biosynthesis. J Cell Biol 99:2092-2098

Goodman DS (1984). Plasma Retinol-binding protein. In Sporn MB,Roberts AB, Goodman DS (eds): "The Retinoids" Orlando, FL: Academic Press, pp 42-82.

Ho M-T P, Massey JB, Pownall HJ, Anderson RE, Hollyfield JG (1989). Mechanisms of vitamin A movement between rod outer segments, interphotoreceptor retinoid-binding protein, and liposomes. J Biol Chem 264:928-935.

Inouye LN, Albini A, Chader GJ, Redmond TM, Nickerson JM (1989). mRNA for interphotoreceptor retinoid-binding protein (IRBP): Distribution and size diversity in vertebrate species. Exp Eye Res 49:171-180.

Johnson AT, Kretzer FL, Hittner HM, Glazebrook PA, Bridges CDB, Lam DMK (1985). Development of the subretinal space in the preterm human eye: ultrastructural and immunocytochemical studies. J Comp Neurol 233:497-505.

Kozak M (1984). Compilation and analysis of sequences upstream from the translational start site in eucaryotic mRNAs. Nucleic Acid Res 12:857-872.

Kyritsis AP, Wiggert B, Lee L, Chader GJ (1985). Butyrate enhances the synthesis of interphotoreceptor retinoid-binding protein (IRBP) by Y-79 human retinoblastoma cells. J Cell Physiol 124:233-239.

Kyte J, Doolittle FR (1982). A simple method for displaying the hydropathic character of a protein. J Mol Biol 157:105-132.

Lai YL, Wiggert B, Liu YP, Chader GJ (1982). Interphotoreceptors retinol-binding proteins: possible transport vehicles between compartments of the retina. Nature 298:848-849.

Laurent BC, Nillson MHL, Bavik CO, Jones TA, Sundelin J, Peterson PA (1985). Characterization of the rat retinol-binding protein gene and its comparison to the three-dimensional structure of the protein. J Biol Chem 260:11476-11480.

Liou GI, Bridges CDB, Fong SL, Alvarez RA, Gonzalez-Fernandez F (1982). Vitamin A transport between retina and pigment epithelium-an interstitial protein carrying endogenous retinol (interstitial retinol-binding protein). Vision Res 22:1457-1467.

Liou GI, Fong SL, Beattie WG, Cook RG, Leone J, Landers RA, Alvarez RA, Wang C, Li Y, Bridges CDB (1986). Bovine interstitial retinol-binding protein (IRBP): isolation and sequence analysis of cDNA clones, characterization and in vitro translation of mRNA. Vision Res 26:1645-1653.

Liou GI, Li Y, Wang C, Fong SL, Bhattacharya S, Bridges CDB (1987a). Bgl II RFLP recognized by a human IRBP cDNA localized to chromosome 10. Nucleic Acids Res 15:3196.

Liou GI, Fong SL, Gosden J, van Tuinen P, Ledbetter DH, Christie S, Rout D, Bhattacharya S, Cook RG, Li Y, Wang C, Bridges CDB (1987b). Human interstitial retinol-binding protein (IRBP): cloning, partial sequence, and chromosomal localization. Somatic Cell and Molecular Genetics 13:315-323.

Liou GI, Ma D-P, Yang Y-W, Geng L, Zhu C, Baehr W (1989). Human Interstitial Retinoid-binding protein: Gene structure and primary sequence. J Biol Chem 264:8200-8206.

Martone RL, Schon EA, Goodman DS, Soprano DR, Herbert J (1988). Retinol-binding protein is synthesized in the mammalian eye. Biochem Biophys Res Commun 157:1078-1084.

Nathans J, Hogness, DS (1984). Isolation and nucleotide sequence of the gene encoding human rhodopsin. Proc Natl Acad Sci USA 81:4851-4855.

Papermaster DS, Dreyer WJ (1974). Rhodopsin content in the outer segment membranes of bovine and frog retinal rods. Biochemistry 13:2438-2444.

Rodrigues MM, Hacket J, Gaskins R, Wiggert B, Lee l, Redmond M, Chader GJ (1986). Interphotoreceptor retinoid-binding protein in retinal rod cells and pineal gland. Invest Ophthalmol Vis Sci 27:844-850.

Saari JC, Futterman S, Bredberg L (1978). Cellular retinol and retinoic-acid binding proteins of bovine retina. Purification and Properties. J Biol Chem 253:6432-6436.

Saari JC, Teller DC, Crabb JW, Bredberg L (1985). Properties of an interphotoreceptor retinoid-binding protein from bovine retina. J Biol Chem 260:195-201.

Schubert D, LaCorbiere M, Esch F (1986). A chick neural

retina adhesion and survival molecule is a retinol-binding protein. J Cell Biol 102:2295-2301.
Si JS, Borst DE, REdmond TM, Nickerson JM (1989). Cloning of cDNAs encoding human interphotoreceptor retinoid-binding protein (IRBP) and comparison with bovine IRBP sequences. Gene 80:99-108.
Stubbs GW, Saari JC, Futterman S (1979). 11-cis-retinal-binding protein from rat retina. Isolation and partial characterization. J Biol Chem 254:8529-8533.
Taniguchi T, Adler AJ, Mizuochi T, Kochibe N, Kobata A (1986). The structures of the asparagine-linked sugar chains of bovine interphotoreceptor retinol-binding protein: occurrence of fucosylated hybrid-type oligosaccharides. J Biol Chem 261:1730-1736.
van Veen T, Katial A, Shinohara T, Barrett DJ, Wiggert B, Chader GJ, Nickerson, JM (1986). Retinal photoreceptor neurons and pinealocytes accumulate mRNA for interphotoreceptor retinoid-binding protein (IRBP). FEBS Letters 208:133-137.
van Veen T, Ekstrom P, Wiggert B, Lee L, Hirose Y, Sanyal S, Chader GJ (1988). A developmental study of interphotoreceptor retinoid-binding protein (IRBP) in single and double homozygous rd and rds mutant mouse retinae. Exp Eye Res 47:291-305.
von Heijne G (1986). A new method for predicting signal sequence cleavage sites. Nucl Acids Res. 14:4683-4690.
Wiggert B, Bergsma D, Lewis M, Chader GJ (1977). Vitamin A receptors: Retinol binding in neural retina and pigment epithelium. J Neurochem 29:947-954.
Wiggert B, Derr JE, Fitzpatrick M, Chader GJ (1979). Vitamin A receptors of the retina: differential binding in light and dark. Biochim Biophys Acta 582:115-121.
Wiggert B, Lee L, Rodrigues MM, Hess HH, Redmond TM, Chader GJ (1986). Immunochemical distribution of interphotoreceptor retinoid-binding protein in selected species. Invest Ophthalmol Vis Sci 27:1041-1049.

Molecular Biology of the Retina: Basic and Clinically Relevant Studies, pages 139–161

THE MOLECULAR BIOLOGY OF IRBP: APPLICATION TO PROBLEMS OF UVEITIS, PROTEIN CHEMISTRY, AND EVOLUTION

John M. Nickerson, Diane E. Borst, T. Michael Redmond, Jing-Sheng Si, Judith Toffenetti, and Gerald J. Chader

LRCMB, NEI, NIH, Bethesda MD 20892

INTRODUCTION

In the visual cycle, 11-cis-retinal, bound by Schiff's base linkage to opsin, is converted by light to all-trans retinol. These retinoids are reused many times, and the all-trans-retinol is apparently converted back to the 11-cis by retinol isomerase (Bernstein, Law and Rando, 1987) and dehydrogenases located in the pigment epithelium. The location of opsin in the photoreceptor and the retinol isomerase in the PE necessitates that the retinoid must cross the interphotoreceptor space (IPS) twice per cycle. Interphotoreceptor retinoid-binding protein (IRBP) has been proposed as a carrier that might facilitate one or both of these crossings (Chader et al., 1983). IRBP is a large glycolipoprotein found in the IPS (Bunt-Milam and Saari, 1983) that is capable of binding retinoids (Wiggert et al., 1977) and fatty acids (Bazan et al., 1985). In the monkey it accounts for about 70% of the protein in this space (Pfeffer et al., 1983). In part because of its abundance, it has been suggested that IRBP might serve as a buffer for retinoids (Ho et al., 1989), perhaps mediating the concentration of free retinoids in the IPS. How it facilitates or mediates is not known but several experiments demonstrating these roles have been performed recently (Jones et al., 1989; Okajima et al., 1989).

We began to study the molecular biology of IRBP by isolating cDNA copies of the bovine retina IRBP mRNA (Barrett et al., 1985). These initial clones have proven useful in their own right, in characterizing the IRBP mRNA and the deduced protein sequence, and also as probes. We have used the bovine cDNA fragments as probes to isolate the human (Nickerson et al., 1988), bovine (Borst and Nickerson, 1988; Borst et al., 1989) and mouse genes (Nickerson et al., 1989) and cDNAs. These clones will be useful for detecting alterations in the gene and mRNA in diseases with a genetic basis. The probes also can be used to detect changes in the amounts of mRNA present and the cellular localization of the mRNA within a tissue (van Veen et al., 1986). This capability could be used to elucidate the temporal and spatial expression of the IRBP gene during development. Another use of the clones is in studying the organization of the genome. The position of the IRBP locus helps to map disease loci and to determine the local environment of the IRBP gene.

IRBP IN DISEASE

Information derived from recombinant DNA approaches can help in the study of eye diseases and complement other approaches already being used. There are many potential diseases, both genetic and nongenetic, for which recombinant DNA techniques can be used. We have applied recombinant DNA techniques to study the autoimmune disorder experimental autoimmune uveitis/pinealitis (EAU/EAP). IRBP purified from bovine retinas induces EAU/EAP in Lewis rats (Gery et al., 1986), monkeys (Hirose et al., 1986), and mice (Caspi et al., 1988). Intact IRBP can cause EAU/EAP with a single dose of just 0.3 μg per rat. Several purified cyanogen bromide fragments of IRBP also cause disease, suggesting that there are several epitopes within the polypeptide that can cause the disease. The IRBP used in these experiments was more than 99% pure, but it is possible that contaminants might cause the disease rather than the IRBP. Two types of experiments rule out this possibility and further characterize the uveitogenic epitopes. The protein sequence of IRBP is shown in Fig. 1. The sequence was deduced from the sequence of the bovine gene (Borst et al., 1989) and from cDNA sequences (Redmond et al., 1989). The first approach makes use of recombinant DNA techniques to produce large parts of the IRBP polypeptide in an expression system, in E. coli.

```
          |---CB15------>|----CB96------------------->
|------------N-TERM----------------->
FQPSLVLEMA QVLLDNYCFP ENLMGMQGAI EQAIKSQEIL SISDPQTLAH VLTAGVQSSL NDPRLVISYE PSTLEAPPRA PAVTNLTLEE IIAGLQDGLR
                                  |----TRP4----------------------> |--TRP8-------->|--------TRP7--------->
        10         20         30         40         50         60         70         80         90        100

                      |-------CB105------->
HEILEGNVGY LRVDDIPGQE VMSKLRSFLV ANVWRKLVNT SALVLDLRHC TGGHVSGIPY VISYLHPGST VSHVDTVYDR PSNTTTEIWT LPEALGEKYS
                                                              |-TRP23>                                |-
       110        120        130        140        150        160        170        180        190        200

                           |-------CB72---------------------------------->
ADKDVVVLTS SRTGGVAEDI AYILKQMRRA IVVGERTVGG ALNLQKLRVG QSDFFLTVPV SRSLGPLGEG SQTWEGSGVL PCVGTPAEQA LEKALAVLML
-TRP6------>|---TRP17---->                       |---TRP13---->
       210        220        230        240        250        260        270        280        290        300

RRALPGVIQR LQEALREYYT LVDRVPALLS HLAAMDLSSV VSEDDLVTKL NAGLQAVSED PRLQVQVVRP KEASSGPEEE AEEPPEAVPE VPEDEAVRRA
                          |---TRP26-------------->                          |---TRP10---------------------> |
       310        320        330        340        350        360        370        380        390        400

LVDSVFQVSV LPGNVGYLRF DSFADASVLE VLGPYILHQV WEPLQDTEHL IMDLRQNPGG PSSAVPLLLS YFQSPDASPV RLFSTYDRRT NITREHFSQT
----TRP1-------------->|------TRP25------>|-----TRP28-------------->                                    |TRP15
       410        420        430        440        450        460        470        480        490        500

                                  |-CB104-->
ELLGRPYGTQ RGVYLLTSHR TATAAEELAF LMQSLGWATL VGEITAGSLL HTHTVSLLET PEGGLALTVP VLTFIDNHGE CWLGGGVVPD AIVLAEEALD
--->         |--TRP5->
       510        520        530        540        550        560        570        580        590        600

                                      |---CB33----->                                         |--------CB43------
RAQEVLEFHR SLGELVEGTG RLLEAHYARP EVVGQMGALL RAKLAQGAYR TAVDLESLAS QLTADLQEMS GDHRLLVFHS PGEMVAEEAP PPPPVVPSPE
            |---TRP7---->|-----TRP26------->
       610        620        630        640        650        660        670        680        690        700

------->
ELSYLIEALF KTEVLPGQLG YLRFDAMAEL ETVKAVGPQL VQLVWQKLVD TAALVVDLRY NPGSYSTAVP LLCSYFFEAE PRRHLYSVFD RATSRVTEVW
                                  |---TRP9----->|--TRP21--->                                             |----
       710        720        730        740        750        760        770        780        790        800

                                                                           |----CB21-------->  |---
TLPHVTGQRY GSHKDLYVLV SHTSGSAAEA FAHTMQDLQR ATIIGEPTAG GALSVGIYQV GSSALYASMP TQMAMSASTG EAWDLAGVEP DITVPMSVAL
-TRP14--->                                                                 |------TRP24----->
       810        820        830        840        850        860        870        880        890        900

---CB41------>                        |---------CB56------------------>
STARDIVTLR AKVPTVLQTA GKLVADNYAS PELGVKMAAE LSGLQSRYAR VTSEAALAEL LQADLQVLSG DPHLKTAHIP EDAKDRIPGI VPMQIPSPEV
                        |------TRP2------>
       910        920        930        940        950        960        970        980        990       1000

                                                           |--------CB58--------------------------->
FEDLIKFSFH TNVLEGNVGY LRFDMFGDCE LLTQVSELLV EHVWKKIVHT DALIVDMRFN IGGPTSSISA LCSYFFDEGP PILLDKIYNR PNNSVSELWT
     |-----TRP3------>
      1010       1020       1030       1040       1050       1060       1070       1080       1090       1100

                                  |--------CB71------------------------->
LSQLEGERYG SKKSMVILTS TLTAGAAEEF TYIMKRLGRA LVIGEVTSGG CQPPQTYHVD DTDLYLTIPT ARSVGAADGS SWEGVGVVPD VAVPAEAALT
                                                                             |-------TRP16--------------->
      1110       1120       1130       1140       1150       1160       1170       1180       1190       1200

RAQEMLQHTP LRARRSPRLH GRRKGHHRQS QGRAGSIGRN QGVVRPEVLT EAPSCQKRCL LQCG
      1210       1220       1230       1240       1250       1260
```

Fig. 1. Bovine IRBP protein sequence deduced from the gene, cDNAs, and direct analysis of the N-terminus, CNBr fragments, and tryptic fragments of the protein. The location of each peptide sequence determined by direct amino acid sequencing is shown by the arrows above the protein sequence.

The expression method has several advantages. First, great quantities of the protein fragment can be obtained. Second, the contaminants of the expressed protein are entirely different from those of bovine eye IRBP preparations. Third, the expression approach is not dependent on perfect sequence determination as a prerequisite to the experiment. Fourth, the polypeptide can be truncated at either end at sites distinctly different from those allowed by the technology of protein chemistry. Finally, the expressed IRBP fragment can constitute as large or small a part of the protein as needed. For example, IRBP contains four 300 amino acid long repeats within the polypeptide, we might choose to study just one or a fraction of one protein repeat.

As a beginning, we fused the N-terminal 584 amino acids of β-galactosidase to the C-terminal 135 amino acids of IRBP (Redmond et al., 1989). The construction is illustrated in Fig. 2.

CONSTRUCTION OF AN EXPRESSION PLASMID THAT PRODUCES A β-GALACTOSIDASE-IRBP FUSION PROTEIN

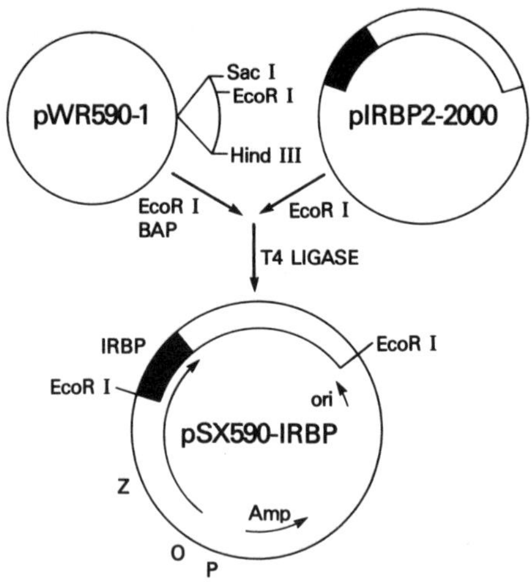

Fig. 2. Construction of an expression plasmid that produces a β-galactosidase-IRBP fusion protein. We inserted a 2 kb EcoR I fragment from λIRBP2 (Barrett et al., 1985) into the EcoR I site of the vector pWR590-2 (Guo et al., 1984). When induced by IPTG, a fusion protein consisting of the N-terminal 584 residues of β-galactosidase, 3 linker residues, and 135 residues from the C-terminal end of IRBP is expressed. The resulting plasmid is called pXS590-IRBP.

We verified that the construction was correct by several means including restriction digestion from which we obtained the appropriate fragments (Fig. 3).

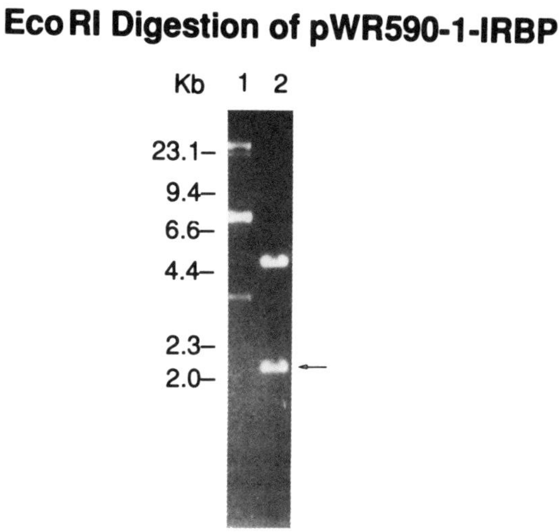

Fig. 3. EcoR I digestion of pXS590-IRBP. Plasmid was isolated (Ish-Horowicz and Burke, 1981) from a culture of cells expressing the IRBP fusion protein. Lane 1 shows undigested supercoiled plasmid. Lane 2 shows plasmid digested with EcoR I. The samples were run on a 0.8% agarose gel containing ethidium bromide. The arrow shows the position of the 2 kb IRBP cDNA insert.

The fusion protein expressed in E. coli was purified by serial detergent/chaotrope extraction of precipitates, Fig. 4, and finally reverse-phase HPLC (Redmond et al., 1989). The β-galactosidase fragment is thought to stabilize the fusion protein, helping to prevent its proteolysis in the bacterium. After purification, the fusion protein caused EAU/EAP in Lewis rats. β-Galactosidase itself does not cause EAU/EAP. This is the first case of a bacterially expressed fusion protein causing an autoimmune disease. These experiments are prerequisites to several types of experiments such as fusing or even chemically cross-linking

Detergent/Chaotrope Extraction of β-gal-IRBP Fusion Protein

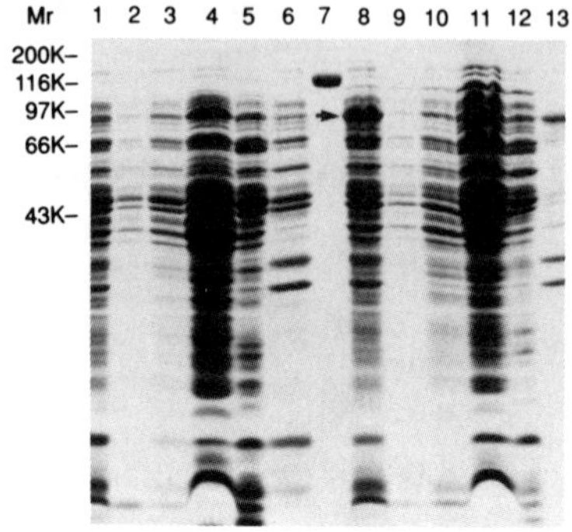

Fig. 4. Detergent/chaotrope extraction (DuBois, 1986) of IRBP fusion protein: SDS-PAGE analysis of samples from intermediate steps in the purification process. Lanes 1-6: LE392 without pXS590-IRBP; Lane 7: authentic bovine IRBP; Lanes 8-13: LE392 with pXS590-IRBP. Lanes 1 and 8: total extracts of E. coli cells; lanes 2 and 9: cell lysate; lanes 3 and 10: 0.05% deoxycholate extract of cell pellet; lanes 4 and 11: 1.5% octyl-β-glucoside extract of pellet from lanes 3 and 10, respectively; lanes 5 and 12: 2 M KSCN extract of pellet from lanes 4 and 11, respectively; lanes 6 and 13; final 2 M KSCN-insoluble pellet. The arrow shows the position of the fusion protein.

IRBP peptides to toxins such as ricin, diphtheria toxin, or others that selectively might kill only the helper T-cells that cause uveitis. Besides possible usefulness in treatment of uveitis, such expressed IRBP fragments might be important in the assessment of IRBP function. The fusion proteins may be useful in testing ligand binding sites: they may be active in binding ligands already. Alternatively, just a part of the peptide might be useful as an antigen for producing antibodies that might inhibit ligand binding in biochemical approaches or in more physiological approaches [see Jones et al. (1989); Okajima et al. (1989)].

The second approach for proving that IRBP causes EAU/EAP is based on synthesis. The strategy is to synthesize a compound and test it for the same/similar properties as one that was purified from a natural source. As with the expression approach, the synthetic material will have very different contaminants from the IRBP preparation obtained from bovine eyes. There are some disadvantages which include: expense, only short oligopeptides are practical, and the technique is dependent on the correct determination of the protein sequence. We provided the protein sequence of IRBP to Igal Gery's and Barbara Wiggert's groups, and they used this information to obtain synthetic oligopeptides. They have tested many oligopeptides based on the DNA derived protein sequence in order to characterize the uveitogenic epitopes (Sanui et al., 1988). One particular oligopeptide, R14, is still potent in doses as low as 60 ng per rat, which on a molar basis is only about five-fold higher than the intact IRBP itself (Sanui et al., 1989). None of the IRBP oligopeptides that cause EAU/EAP is similar in primary structure to S-antigen, the other well-characterized highly uveitogenic protein of the eye (Shinohara et al., 1988).

ANALYSIS OF IRBP mRNA

An analysis of IRBP mRNA is necessary for an understanding of IRBP gene expression. By retinoid-binding (Carter-Dawson et al., 1986) and immunoassay (van Veen et al., 1988) IRBP appears in mice at about the time of rod outer segment elongation. This raises questions about the possible coordination of IRBP expression and outer segment development. The neural retina is the site of synthesis of IRBP (Hollyfield et al., 1985; Wiggert et al., 1984). By <u>in situ</u> hybridization, IRBP mRNA accumulates only in the perikaryon of the photoreceptor. This defines the precise cell in which the gene is transcribed actively and thus the only retina cell which can translate the mRNA into protein (van Veen et al., 1986).

We have estimated the abundance of IRBP mRNA from its frequency in a human retina cDNA library. The abundance is 1 per 800 clones (Si et al., 1989). Assuming the frequency has not been biased by cloning and amplification, it suggests a similar frequency among retina mRNAs. We have examined retina RNAs from several mammalian and bird species (Inouye et al., 1989). We find one principal or just a single band by northern RNA blot analysis. There is a large variation in size, but all species examined so far are more than 4.3 kb. All are long enough to encode a polypeptide of 130-140,000 MW. The largest mRNA is hamster at 7.6 kb. The major source of variability among species may be in the 3' untranslated region. This is the case in the comparison of the human and bovine mRNAs (Si et al., 1989). Fig. 5 shows a dot matrix comparison of the bovine and human cDNA sequences.

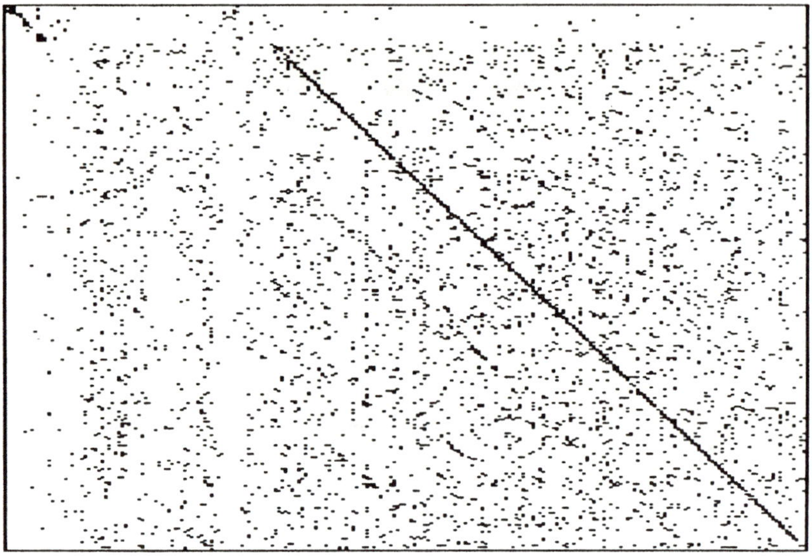

Fig. 5. Dot matrix comparison (Maizel and Lenk, 1981) of the human and bovine cDNA sequences of IRBP. The bovine sequence is on the abscissa and the human sequence is on the ordinate. The DIAGON program (Staden, 1984) was used with a SPAN of 13 and a score of 10 using the identity algorithm. The analysis shows a major insertion into the bovine sequence or a major deletion in the human sequence in the 3′ untranslated regions. The diagonal line in the lower right corner of the matrix shows that the two cDNAs retain sequence similarity in the last 270 bases of the cDNAs. Weak and badly broken lines of homology parallel to major diagonal can be detected by tilting the figure and looking along the major axis. These weak lines show the repeat structure which is much more obvious at the protein level.

The human cDNA sequence determination is described later and in Si et al. (1989). The sequences are very similar as indicated by the diagonal line, but there is a 1.58 kb insertion/deletion in the 3′ untranslated region of each of the mRNAs located about 275 bases from the poly (A) tail. We find mRNA for IRBP in the retina and the pineal (Fig. 6), but in the pineal it is much less abundant, very roughly 25-fold less (Toffenetti et al., 1989).

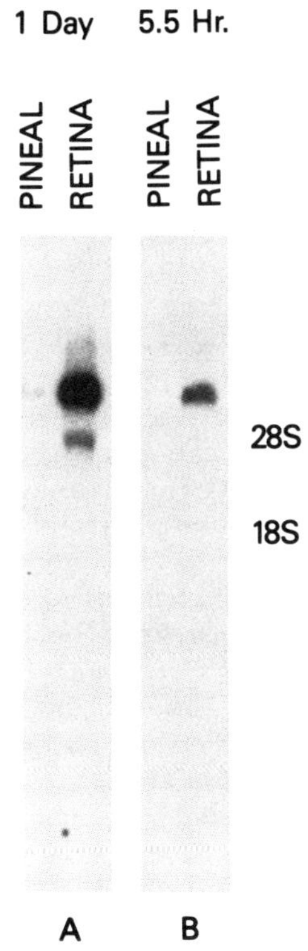

Fig. 6. Northern blot analysis of total retina and pineal RNA. Two exposures of the same blot are shown (Panel A is a 1 day exposure and Panel B is a 5.5 h exposure). The probe was from a bovine IRBP cDNA.

The mRNA is the same size and just a single principal species in bovine pineal. It would be reasonable to expect that the vestigial nature of the photoreceptor-like functions in the pineal contributes to the lower level of IRBP mRNA in the pineal compared to the retina. The differential levels of IRBP mRNA in these two tissues suggests that there are regulatory mechanisms that allow IRBP mRNA levels to be set. We are investigating these mechanisms now.

IRBP GENE ANALYSIS

The number of copies of the IRBP gene per haploid genome in the cow and several other animals is probably only one (Borst and Nickerson, 1988). This was determined by inspection of the number and sizes of bands on genomic blots of bovine DNA digested individually with several different restriction enzymes. The sizes of the bands added up to a remarkably small size for the IRBP gene. We obtained similar results from several different species. This also demonstrates an evolutionary conservation at the DNA level. We could even detect IRBP-like gene fragments in three birds (chicken, duck and turkey) that we tested. There were barely detectable levels of putative IRBP mRNA in chicken, duck and quail retina RNAs (Inouye et al., 1989), but again the sizes were consistent with protein sizes of $\geq$ 135,000 MW. These experiments established conditions under which we could examine the IRBP genes in different species with interesting eye mutations. The conditions also allow investigations of the gene size and structure in other species.

IRBP concentrations are low in eyes from individuals with ARRP (Rodrigues et al., 1986), ADRP (Rodrigues et al., 1985) and also choroideremia (Rodrigues et al., 1984), suggesting that IRBP may be involved in the pathogenesis of these diseases. The question arose: does the IRBP gene cause the disease in any of these disorders? Because IRBP was lost selectively in a young individual with choroideremia, a X-linked disorder, it seemed plausible that IRBP was the genetic basis of the disease. We tested this possibility with the newly developed genomic blotting conditions. We compared the band intensities of human male and female DNA and found the intensities to be the same, suggesting that the IRBP gene is on an autosome, not associated with the X-chromosome where choroideremia maps (Borst and Nickerson, 1988). Again this proves the usefulness of applying molecular biology approaches to clinically significant questions. Other groups and we have followed up on the determination of the chromosomal localization of the IRBP gene and found it near the centromere of chromosome 10. We found it near 10q21.1 by in situ hybridization (Nickerson et al., 1988). Collaborators found it in the same spot (Ngo-

Jones et al., 1989) by in situ and confirmed with panels of rodent-human hybrid cells. Further confirmation was obtained by the isolation of a genomic clone and its sequence analysis from a human chromosome 10-specific library. Others (Liou et al., 1987) have localized the IRBP gene to 10p11.2 to 10q11.2. They find, near the centromere, that more grains (13) are on the short arm than on the long arm (9), and find a secondary site on chromosome 10 at q24-q25.

GENE STRUCTURE

We determined the nucleotide sequence of the bovine IRBP gene (Borst et al., 1989). The gene is 11.6 kb long and contains 4 exons and 3 introns. There is a four-fold repeat in the predicted protein sequence, with 30-40% identity between any two repeats. The third and fourth repeats are the most similar. Each repeat is about 300 amino acids long. All 3 introns lie in that part of the gene encoding the fourth protein repeat. Each intron is 1.5 to 2.2 kb in length. The fourth exon is 2.4 kb long and includes 2.0 kb of 3′ untranslated region. The first exon is among the longest discovered yet, 3173 bases long. It encodes 1016 amino acids and more than three full protein repeats. It includes putative signal and pro-peptides. By optimal alignment of the 4 protein repeats 42 residues are identical among all four of the repeats and in 62 residues 3 of the 4 match. Some of the residues are clustered suggesting that these locations in the sequences may be particularly important in maintaining tertiary structure or in ligand/receptor binding. Other features align among the repeats including possible phosphorylation sites and hydrophobic areas, and glycosylation sites. For example, 3 of the 5 N-linked carbohydrate sites align among the 4 repeats.

We also have determined the sequence of a full length cDNA for human IRBP mRNA, and the sequence of a human gene clone which originated from a chromosome 10-specific genomic library. The sequencing strategy of the three human cDNA clones λIRBP18-3500, λIRBP19-1000, and λIRBP20-700 is shown in Fig. 7.

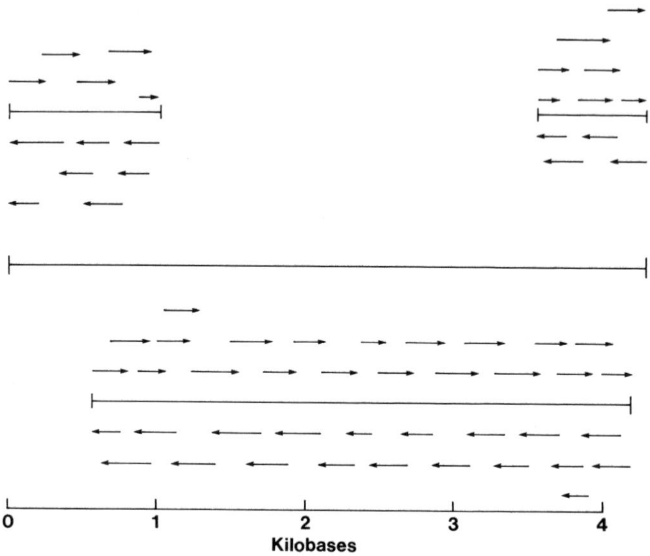

Fig. 7. Sequencing strategy for the three clones comprising the complete cDNA sequence for human IRBP. Sequences from the ends of EcoR I fragments were obtained by conventional dideoxy-sequencing techniques from single-stranded DNAs generated from phagemids and the universal primer. Later primers were based on the prior runs or on a 2.2 kb cDNA sequence (Liou et al., 1987). Each strand of each clone is continuous.

A comparison of the human and bovine protein sequences by dot matrix analysis is shown in Fig. 8. The four-fold repeat structure is obvious. Each repeat is 80-90% identical cross-species. Intraspecies, among the four repeats there is 30-40% identity. But the third and fourth repeats are slightly more similar than other pairs. There is conservation of the position of glycosylation sites among the repeats. Within the eight repeats characterized so far (4 from bovine and 4 from human) there are a total of 7 putative Asn-linked carbohydrate sites. Five of the seven are positioned identically within the aligned repeats.

In our comparison of the bovine gene and the human chromosome 10 IRBP gene fragment we found that the boundary

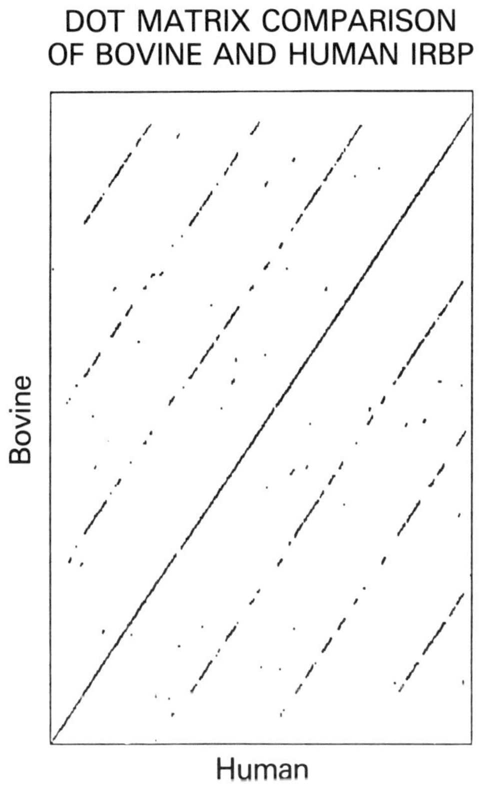

Fig. 8. Dot matrix comparison of the human and bovine IRBP polypeptides. The human polypeptide sequence is on the abscissa and the bovine is on the ordinate. The DIAGON program (Staden, 1984) was used with the proportional algorithm. The SPAN is 15 and the SCORE is 180. The major diagonal line shows the similarity of the bovine and human sequences. Along any column or row, there are four diagonal lines, each one corresponding to about 300 amino acids, showing a four fold repeat structure within the polypeptide. The similarity between any one of the bovine repeats and any of the human repeats is evident.

---

of Intron B and Exon 3 are identical in these species. Liou et al. (1989) also find splice sites in the same locations.

A comparison of the bovine and human signal and pro-peptides is shown in Fig. 9. Each unprocessed polypeptide contains a putative signal peptide of 17 amino acids and putative propeptides of 5 amino acids.

```
HUMAN UNPROCESSED  MMREWVLLMSVLLCGLAGPTHLFQPSLVL.....
HUMAN "N+5"                         GPTHLFQPSLVL.....
HUMAN "N"                                FQPSLVL.....
BOVINE UNPROCESSED MVRKWALLLPMLLCGLTGPAHLFQPSLVL.....
BOVINE MATURE                            FQPSLVL.....
                                         |
                   |<---SIGNAL---->|<-->||
                        PEPTIDE     PRO- |
                                 PEPTIDE |
                                         N-TERMINUS
```

Fig. 9. Alignment of human and bovine IRBP sequences before and after post-translational processing. The putative primary translation product without any type of processing is labeled "UNPROCESSED." The sequence of secreted bovine IRBP as determined by Edman degradation is labeled "MATURE." Two sequences of human IRBP determined by Edman degradation are labeled "N" and "N+5." Two populations of human IRBP exist in the IPM. They differ only in the presence (N+5) or absence (N) of a 5 amino acid extension on the N-terminal end of the human polypeptide (Redmond et al., 1986). The putative signal peptides and propeptides are discussed in Borst and Nickerson (1988).

The human sequence begins with two Met codons in a row. About 70% of the residues in the signal peptides are identical, and many of the remaining residues are conserved. The human mature IRBP obtained from the IPM is a population of polypeptides, some of which begin at position -5 and about 50% that begin at position +1. The bovine protein isolated from the IPM begins only at position +1. The five residues, GPTHL, in human and GPAHL in bovine we call the propeptide. Its function if any is not known.

EVOLUTION OF THE IRBP GENE

All vertebrates which have been examined seem to have IRBP. This suggests that IRBP arose 0.5 billion or more years ago. Few invertebrates have been tested except for squid. A

48,000 MW squid retinaldehyde binding protein (Fong et al., 1988) has been found which cross-reacts with anti-bovine IRBP. This protein may represent a monomeric relative of the IRBP gene. Thus, it seems reasonable to suppose that quadruplicated IRBP arose about the time of the radiation of the vertebrates. What advantages would an IRBP polypeptide have if it had four repeats rather than one? If one or more of the repeats became inactive, the protein could still retain at least some function(s) via its other repeats. This might have occurred already in bovine IRBP since it binds only one or two retinoids per polypeptide instead of up to as many as four. It could free other repeats to bind other types of ligands or to perform other functions. IRBP does bind noncovalently up to 4 mol of fatty acids per mol of polypeptide. We might speculate that one might bind to each repeat. Similarly, cooperativity among repeats in binding/releasing ligands might be very important for vision, especially in physiologically demanding circumstances such as after intense light-bleaching, disc shedding, or during dark adaptation. Any of these possibilities could provide sufficient disadvantage for a monomeric protein versus its quadruplicated counterpart as ancestral species entered new environments and evolutionary niches.

The unexpectedly small gene size, the very large first exon, and the 4 repeat protein structure lead us to propose three models for the evolution of the IRBP gene. These models tend to emphasize extremes, and we do not intend to be exhaustive in enumerating possible models. Nor do we intend to rule out the possibility of combining features from two or more models. The first model, Fig. 10, is an intron insertion model. Starting with an ancestral monomeric gene with no introns, two rounds of internal duplications could generate a quadruplication. Three introns could be inserted in the fourth repeat to produce a gene structure similar to the bovine gene. There are several problems with this model which make it seem less likely than the third model that we present. First, intron insertion is rare. Second, only in one of 18 occasions would the 3 introns fall in the last protein repeat. Third, internal duplications are rare in comparison to mechanisms such as unequal crossovers; here we invoked two rounds of duplication.

A second possibility, Fig. 11, begins with an ancestral 300 amino acid monomeric gene that already contains 3 introns.

Again 2 serial rounds of duplication produce a quadruplication. The major complication of this model is

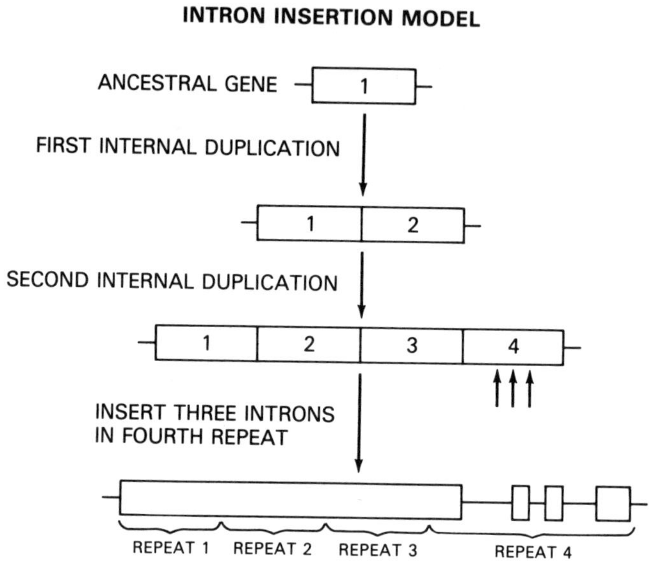

Fig. 10. A model for the evolution of the IRBP gene, based on the insertion of introns. A primordial single exon gene encoding a protein of about 300 amino acids could serve as the ancestral IRBP gene. By two serial internal gene duplications a quadruplicated protein could emerge. Following the insertion of 3 introns into the part of the gene encoding the fourth protein repeat, a gene resembling the present day IRBP gene is obtained.

the loss of introns A-I (the first nine introns). While intron loss occurs with a measurable frequency, it seems unlikely that it could occur on 9 separate occasions and on only the first 9 introns. This model still results in a gene that strongly resembles the current-day bovine IRBP gene. If this intermediate were to undergo a partial gene conversion with an IRBP cDNA to remove introns A-I, it would result in a more reasonable model.

A third possibility, Fig. 12, begins as the second, with an ancestral 300 amino acid protein encoded by a gene with 3 introns.

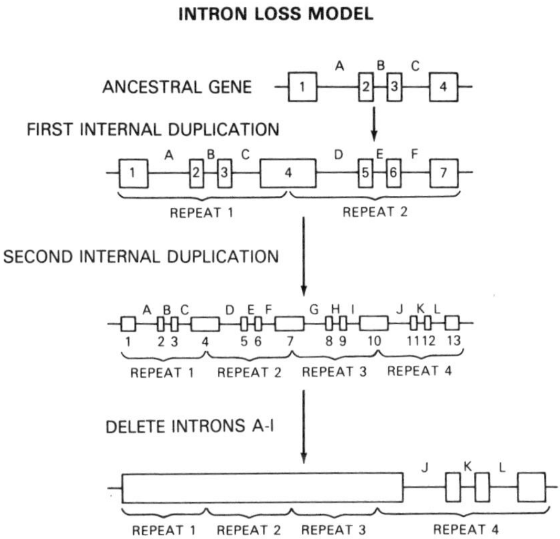

Fig. 11. A second model for the evolution of the IRBP gene. Here the ancestral monomeric gene already might have 3 introns in it. Two serial internal duplications generate a quadruplicated protein, but with 12 introns in it. If the first 9 introns were deleted we could obtain a gene with a structure similar to the current IRBP gene.

A processed gene, possibly derived from a viral reverse transcription of IRBP mRNA, is inserted at the beginning of the ancestral gene (which ultimately will become the beginning of the fourth protein repeat). This insertion generates a two repeat gene. This event is the only one expected to be rare, although no more rare than the straightforward duplication invoked in the previous two models. Two unequal crossover events generate the quadruplication. Neither of these two crossover events can be considered rare. The result is a gene that looks very much like the present mammalian IRBP gene. The use of a

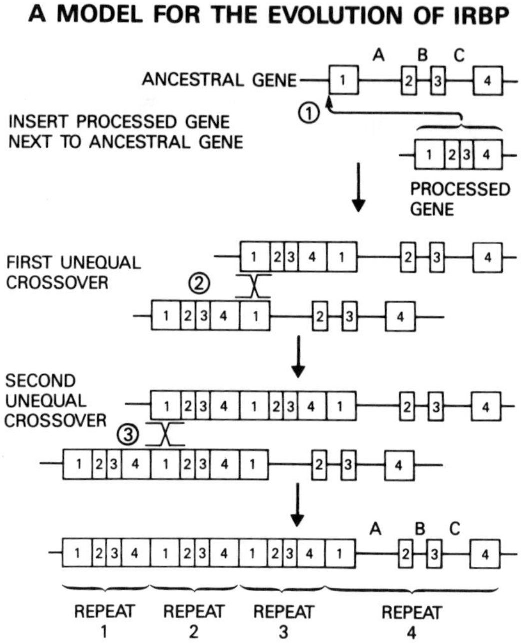

Fig. 12. A model for the evolution of the bovine IRBP gene. The intent is to offer a possible model that explains the absence of introns in the first 1016 amino acids of the gene. In this model, we propose that an ancestral gene of about 300 amino acids contained 3 introns labeled A, B, and C in the figure. The exons in the putative ancestral gene are labeled 1, 2, 3 and 4. The first step in the evolution of the IRBP gene is the insertion of a processed gene in the 5′ end of the ancestral gene. This generates a gene intermediate with an internal duplication. Two unequal crossovers subsequently can generate a fourfold repeat in the IRBP gene with just 4 exons, the first exon being very large and encoding 3 full repeats and part of the fourth repeat.

processed gene cannot be considered unusual, many genes, > 200, have processed pseudogenes. There are several examples of processed genes inserting near their parental gene (Hardison et al., 1986; Nishioka et al., 1980). In many ways the third model seems the most plausible.

SUMMARY

In closing, molecular biology has helped us in the study of uveitis, in defining the parts of the IRBP molecule that cause EAU/EAP. Simple northern and Southern blotting experiments can answer significant questions in the study of gene structure and expression and in relating these to genetic eye diseases. Molecular biology has provided answers to several long-standing cell biology questions such as: where is IRBP synthesized, where is the gene expressed, and how is the IRBP polypeptide processed? In considering how mother nature invented the IRBP gene, the gene structure suggests some interesting alternative models, and causes us to speculate about how large proteins may have evolved. Finally, the determination of the protein sequence helps to put us in the position to ask and answer questions about the function of IRBP. It also allows us to begin to determine the consequences of a nonfunctional IRBP. We don't know the answers to all these questions yet, but the structural analyses and the isolation of these genes and cDNAs, presented here, united with other powerful biological techniques should provide the answers.

REFERENCES

Barrett DJ, Redmond TM, Wiggert B, Oprian DD, Chader GJ and Nickerson JM (1985). cDNA clones encoding bovine interphotoreceptor retinoid binding protein. Biochem Biophys Res Commun 131:1086-93.

Bazan NG, Reddy TS, Redmond TM, Wiggert B and Chader GJ (1985). Endogenous fatty acids are covalently and noncovalently bound to interphotoreceptor retinoid-binding protein in the monkey retina. J Biol Chem 260: 13677-13680.

Bernstein PS, Law WC and Rando RR (1987). Isomerization of all-trans-retinoids to 11-cis-retinoids in vitro. Proc. Nat'l. Acad Sci. (USA) 84:1849-1853.

Borst DE and Nickerson JM (1988). The isolation of a gene encoding interphotoreceptor retinoid-binding protein. Exp. Eye Res. 47:825-838.

Borst DE, Redmond TM, Elser JE, Gonda MA, Wiggert B, Chader GJ and Nickerson JM (1989). Interphotoreceptor retinoid binding protein: gene characterization, protein repeat structure, and it evolution. J. Biol. Chem. 264:1115-1123.

Bunt-Milam AH and Saari JC (1983). Immunocytochemical localization of two retinoid-binding proteins in vertebrate retina. J Cell Biol 97:703-712.

Carter-Dawson L, Alvarez RA, Fong SL, Liou GI, Sperling HG and Bridges CD (1986). Rhodopsin, 11-cis vitamin A, and interstitial retinol-binding protein (IRBP) during retinal development in normal and rd mutant mice. Dev Biol 116:431-438.

Caspi RR, Roberge FG, Chan CC, Wiggert B, Chader GJ, Rozenszajn LA, Lando Z and Nussenblatt RB (1988). A new model of autoimmune disease. Experimental autoimmune uveoretinitis induced in mice with two different retinal antigens. J Immunol 140:1490-1495.

Chader GJ, Wiggert B, Lai Y-L and Fletcher R (1983). Interphotoreceptor retinoid-binding protein: a possible role in retinoid transport to the retina. Prog. Ret. Res. 2:163-189.

DuBois GC (1986). Rapid purification of bacterially expressed fusion proteins by HPLC methods. Gene Anal. Techn. 3:6-11.

Fong S-L, Lee PG, Ozaki K, Hara R, Hara T and Bridges CDB (1988). IRBP-like proteins in the eyes of six cephalopod species--immunochemical relationship to vertebrate interstitial retinol-binding protein (IRBP) and cephalopod retinal-binding protein. Vision Res 28: 563-573.

Gery I, Wiggert B, Redmond TM, Kuwabara T, Crawford MA, Vistica BP and Chader GJ (1986). Uveoretinitis and pinealitis induced by immunization with interphotoreceptor retinoid-binding protein. Invest Ophthalmol Vis Sci 27:1296-1300.

Guo L-H, Stepien PP, Tso JY, Brosseau R, Narang S, Thomas DY and Wu R (1984). Synthesis of human insulin gene VIII. Construction of expression vectors for fused proinsulin production in E. coli. Gene 29:251-254.

Hardison RC, Sawada I, Cheng JF, Shen CK and Schmid CW (1986). A previously undetected pseudogene in the human alpha globin gene cluster. Nucl Acids Res 14: 1903-1911.

Hirose S, Kuwabara T, Nussenblatt RB, Wiggert B, Redmond TM and Gery I (1986). Uveitis induced in primates by interphotoreceptor retinoid-binding protein. Arch Ophthalmol 104:1698-1602.

Ho M-TP, Massey JB, Pownall HJ, Anderson RE and Hollyfield JG (1989). Mechanism of vitamin A movement between rod outer segments, interphotoreceptor retinoid-binding protein, and liposomes. J. Biol. Chem. 264:928-935.

Hollyfield JG, Fliesler SJ, Rayborn ME, Fong S-L, Landers RA and Bridges CD (1985). Synthesis and secretion of interstitial retinol-binding protein by the human retina. Invest. Ophthalmol. Vis. Sci. 26:58-67.

Inouye LN, Albini A, Chader GJ, Redmond TM and Nickerson JM (1989). mRNA for interphotoreceptor retinoid-binding protein (IRBP): distribution and size diversity in vertebrate species. Exp. Eye Res. 49:171-180.

Ish-Horowicz D and Burke JF (1981). Rapid and efficient cosmid cloning. Nucl. Acids Res. 9:2989-2998.

Jones GJ, Wiggert B, Crouch RK, Cornwall MC and Chader GJ (1989). Bovine IRBP and amphibian photoreceptors: retinoid transfer and protection. Invest Ophthalmol Vis Sci (Suppl) 30:488.

Liou GI, Fong SL, Gosden J, van Tuinen P, Ledbetter DH, Christie S, Rout D and Bhattacharya S (1987). Human interstitial retinol-binding protein (IRBP): cloning, partial sequence, and chromosomal localization. Somat Cell Mol Genet 13:315-323.

Liou GI, Ma D-P, Yang Y-W, Li G, Cihui Z and Baehr W (1989). Human interstitial retinoid-binding protein: gene structure and primary sequence. J Biol Chem 264:8200-8206.

Maizel JV, Jr and Lenk RP (1981). Enhanced graphic matrix analysis of nucleic acid and protein sequences. Proc Natl Acad Sci USA 78:7665-9.

Ngo-Jones JT, Klisak I, Nickerson JM, Borst DE, Redmond TM, Mohandas T, Sparkes RS, Barrett DJ and Bateman JB (1989). Assignment of an interphotoreceptor retinoid-binding protein (IRBP) gene to human chromosome 10. (submitted).

Nickerson JM, Borst DE, Chader GJ and Si J-S (1989). Evolution of the IRBP gene: Comparison of the human, bovine, and mouse sequences. Invest Ophthalmol Vis Sci (Suppl) 30:115.

Nickerson JM, Borst DE, Redmond TM, Hershfield B, Si J-S, Albini A, Inouye L and Toffenetti J (1988). Molecular Genetics of Interphotoreceptor Retinoid-binding Protein. In Piatigarsky J, Shinohara T and Zelenka PS. (Eds.) Molecular Biology of the Eye: Genes, Vision, and Ocular Diseases. Alan R. Liss, Inc. New York. pp. 93-105.

Nishioka Y, Leder A and Leder P (1980). Unusual alpha-globin-like gene that has cleanly lost both globin intervening sequences. Proc Natl Acad Sci (USA) 77: 2806-2809.

Okajima T-IL, Pepperberg DR, Ripps H, Wiggert B and Chader GJ (1989). Interphotoreceptor retinoid-binding protein: role in the delivery of retinol to the pigment epithelium. Exp. Eye Res. (in press).

Pfeffer B, Wiggert B, Lee L, Zonnenberg B, Newsome D and Chader G (1983). The presence of a soluble interphotoreceptor retinol-binding protein (IRBP) in the retinal interphotoreceptor space. J Cell Physiol 117:333-341.

Redmond TM, Wiggert B, Robey FA and Chader GJ (1986). Interspecies conservation of structure of interphotoreceptor retinoid-binding protein. Similarities and differences as adjudged by peptide mapping and N-terminal sequencing [Erratum Biochem J 242 (1987) 935]. Biochem J 240:19-26.

Redmond TM, Si J-S, Barrett DJ, Borst DE, Rainier S, Kotake S, Gery I, and Nickerson J (1989). Synthesis of an immunopathogenic fusion protein derived from a bovine interphotoreceptor retinoid-binding protein cDNA clone. Gene 80:109-117.

Rodrigues MM, Ballintine EJ, Wiggert BN, Lee L, Fletcher RT and Chader GJ (1984). Choroideremia: a clinical, electron microscopic, and biochemical report. Ophthalmology 91:873-883.

Rodrigues MM, Wiggert B, Hackett J, Lee L, Fletcher RT and Chader GJ (1985). Dominantly inherited retinitis pigmentosa: ultrastructure and biochemical analysis. Ophthalmology 92:1165-1172.

Rodrigues MM, Wiggert B, T'so MO and Chader GJ (1986). Retinitis pigmentosa: immunohistochemical and biochemical studies of the retina. Can J Ophthalmol 21: 79-83.

Sanui H, Redmond TM, Hu L-H, Kuwabara T, Margalit H, Cornette JL, Wiggert B, Chader GJ and Gery I (1988). Synthetic peptides derived from IRBP induce EAU and EAP in Lewis rats. Curr. Eye Res. 7:727-735.

Sanui H, Redmond, TM, Kotake S, Wiggert B, Hu L-H, Margalit H, Berzofsky JA, Chader GJ, and Gery, I. (1989). Identification of an immunodominant and highly immunopathogenic determinant in the interphotoreceptor retinoid-binding protein (IRBP). J. Exp. Med. 169:1947-1960.

Shinohara T, Donoso L, Tsuda M, Yamaki K and Singh VK (1988). S-Antigen: structure, function and experimental autoimmune uveitis (EAU). Prog. Retinal Res. 8:51-65.

Si J-S, Borst DE, Redmond TM and Nickerson JM (1989). Cloning of cDNAs encoding human interphotoreceptor retinoid-binding protein (IRBP) and comparison with bovine IRBP sequences. Gene 80:99-108.

Staden R (1984). Graphic methods to determine the function of nucleic acid sequences. Nucleic Acids Res 12:521-538.

Toffenetti J, Chader GJ and Nickerson JM (1989). Analysis of possible variations within cDNAs of bovine interphotoreceptor retinoid-binding protein (IRBP): validation of putative cDNA clones using the polymerase chain reaction (PCR). (submitted).

van Veen T, Ekstrom P, Wiggert B, Lee L, Hirose Y, Sanyal S and Chader GJ (1988). A developmental study of interphotoreceptor retinoid-binding protein (IRBP) in single and double homozygous rd and rds mutant mouse retinas. Exp Eye Res 47:291-305.

van Veen T, Katial A, Shinohara T, Barrett DJ, Wiggert B, Chader GJ and Nickerson JM (1986). Retinal photoreceptor neurons and pinealocytes accumulate mRNA for interphotoreceptor retinoid-binding protein (IRBP). FEBS Lett 208:133-137.

Wiggert B, Bergsma DR, Lewis M and Chader GJ (1977). Vitamin A receptors: retinol binding in neural retina and pigment epithelium. J Neurochem 29:947-954.

Wiggert B, Lee L, O'Brien PJ and Chader GJ (1984). Synthesis of interphotoreceptor retinoid-binding protein (IRBP) by retina in organ culture: effect of monensin. Biochem Biophys Res Commun 118:789-796.

Molecular Biology of the Retina: Basic and Clinically Relevant Studies, pages 163–190

S-ANTIGEN: MOLECULAR MIMICRY MAY PLAY A ROLE IN AUTOIMMUNE UVEITIS

Shinohara T, Singh VK, Yamaki K, Abe T, Tsuda M and Suzuki S.

Molecular Biology Section, Laboratory of Retinal Cell and Molecular Biology, National Eye Institute, National Institutes of Health, Bethesda, Maryland 20892

1. INTRODUCTION

Intraocular inflammatory conditions often classified as "uveitis" are a major cause of visual impairment. They affect approximately 10% of visually impaired patients in the USA and the world. Although the etiology is unknown, T-cell mediated autoimmune processes are thought to play a major role in the pathogenesis of some forms of human uveitis (see Faure, 1980. Gery et al., 1986).

The pathogenesis of ocular autoimmune infla-mmtory diseases has been investigated since early in this century (see Faure, 1980). Wacker and Lipton (1965) showed that retinal tissue emulsified in complete Freund's adjuvant (CFA) is a highly efficient immunogen for the induction of experimen-tal autoimmune uveitis(EAU). S-Antigen (S-Ag), a potent uveitopathogenic antigen from retina, was later isolated in highly purified form (Wacker et al., 1977. Dory and Farue, 1977). This protein is also called as the 48 KDa-protein, an abundant photoreceptor protein important in visual transduc-tion (Kuhn et al., 1984. Pfister et al., 1985). However, the molecular weight of this hydrophilic protein has been shown to be actually 45 kilodal-tons (Yamaki et al., 1987. Shinohara et al.,1987). It is also present in the pineal gland (Kalsow and

Wacker, 1977).

Recently, we determined the entire amino acid sequence of S-Ag and identified its uveitopathogenic sites. Multiple peptides corresponding to specific regions of the amino acid sequence of bovine S-Ag have been synthesized and found to be potent agents for the induction of EAU (Donoso et al., 1986, 1987a and 1987b, Singh et al., 1988a, 1988b and 1989a) and experimental autoimmune pinealitis (EAP) (Kalsow and Wacker, 1978) in laboratory animals.

We can also induce EAU and pinealitis in Lewis rats with a small synthetic peptide corresponding to a certain sequence in baker's yeast histone H3 which contains five consecutive amino acid residues identical to one of the uveitopathogenic sites in human S-Ag (Singh et al., 1989c and 1989d). Thus, molecular mimicry, a situation where a host protein shares the same epitope with a microbe protein, is important in the etiology of autoimmune uveitis.

In this report, we summarize the primary structure and uveitopathogenic sites in S-Ag, and the possibility of molecular mimicry between S-Ag and foreign proteins. These findings and other autoimmune diseases such as EAE will be discussed with respect to the etiology of autoimmune inflammatory uveitis.

2. MOLECULAR MIMICRY

Although many theories have been proposed for the induction of autoimmunity, the etiologies of most of these conditions are unknown. One mechanism for the development of autoimmune uveitis involves "molecular mimicry", whereby selected antigenic determinants from viruses and bacteria can elicit the formation of antibodies or effector lymphocytes, which in turn may react with homologous epitopes of a host antigen and result in the destruction of "self" tissues. This process does not require either an ongoing replicating agent or the continuous presence of the immunogen, i.e. the immunologically mediated injury can occur after the removal of the original immunogen. An ongoing cycle could thus be initiated and the autoimmune response would lead to tissue injury that, in turn, could release more self-antigen inducing an additional lymphocyte response (Fujinami and Oldstone, 1985. Oldstone, 1987. Cohen, 1988).

In theory autoimmunity provoked by molecular mimicry should occur only when nonself- and self-determinants are similar enough to cross-react, yet different enough to break immunological tolerance. The induction and breaking of tolerance at both B-cell and T-cell levels have been established in heterogeneous serum protein models (Oldstone, 1987. Dyrberg, 1986. Oldstone and Notkins, 1986).

T-cells do not recognize the intact protein molecule but rather a small peptide (Watt et al., 1984). The recognition of the epitope by T cells requires a trimolecular complex consisting of a class I or class II MHC molecule, the T cell antigen receptor and the antigen (see Schwartz, 1985).

3. EXPERIMENTAL ANIMAL MODELS

The pathogenicity of autoimmune diseases in the human cannot be studied easily, therefore appropriate animal models offer some insight into understanding the mechanisms of autoimmunity. EAU is one such model for studying certain human uveitis conditions. Some of the models have been well established during last 2-3 decades. The EAU can be induced in a variety of species (guinea pig, rat and monkey) by sensitization with either whole retina or retinal proteins such as S-Ag (Wacker et al., 1977), interphotoreceptor retinal binding protein (IRBP) (Wiggert et al., 1985) and rhodopsin (see Gery et al., 1986) with CFA. EAU is characterized by an inflammation of uvea and retina with loss of photoreceptor cell function (Faure, 1980. Gery et al., 1986)

Although, humoral antibodies (B-cells) may play a role in the development of EAU,(de Kozak et al., 1976 and 1979), it has been characterized predominantly as a T-cell (helper/inducer) mediated autoimmune disease of the uveal tract and the retina (Mochizuki et al., 1984 and 1985, Chan et al., 1985. Caspi et al .,1986). T helper/inducer cells play similar roles in several other experimental autoimmune animal models including EAE, EAT, EAN, AA and CA (table 1) (Kumar et al., 1989). Cytotoxic T-cells are believed to be the delayed

Table 1. A summary of the experimental animal models of autoimmune inflammatory diseases.

Animal models	Relevant human diseases	Self Antigens	Primary mediators
EAU	Uveitis	S-Ag & IRBP	T cells
Experimental autoimmune encephalo-myelitis (EAE)	Multiple sclerosis	Myelin basic protein	T cells
Experimental autoimmune thyroiditis(EAT)	Hashimoto thyroiditis	thyroid-globulin	T cells
Experimental autoimmune neuritis (EAN)	Guillian Barre syndrome	P2 protein Peripheral nervous	T cells
Adjuvant arthritis (AA)	Rheumatoid arthritis	Proteoglycan	T cells
Collagen induced arthritis (CA)	Rheumatoid arthritis	Type 11 collagen	T cells

References, EAE: (Bottazzo et al., 1986. Kumar et al., 1989). EAT: (Bottazzo et al., 1986. Wick et al., 1986. Kong et al., 1986. Cohen, 1986). EAN: (Linington et al., 1984). AA: (Shoenfeld and Isenberg, 1988). CA: (Kakimoto et al., 1988)

type hypersensitivity (DTH) in EAU (Mochizuki et al., 1985) and in other systems (see Kumar et al., 1989).

The histology of EAU induced by S-Ag in primates is similar to some types of human uveitis such as sympathetic ophthalmia, birdshot retino-choroidopathy,Vogt-Koyanagi-Harada syndrome and Behcet's disease (Nussenblatt et al., 1981). EAU induced by native S-Ag,(Nussenblatt et al., 1980) and peptide M (Hirose et al., 1989) are also found to stimulate proliferative responses in lymphocyte cultures of some patients with uveitis (Shinohara et al., 1988b). These results suggest that EAU is a true and accurate model for some types of human uveitis.

4. STRUCTURE OF S-AG

It is essential to know the structure of specific host proteins to be able to study molecular mimicry. Recently, the entire amino acid sequence of human (Yamaki et al., 1988), bovine (Shinohara et al., 1987. Yamaki et al., 1987), and mouse (Tsuda et al., 1988) retinal S-Ag as well as rat pineal gland (Abe et al., 1989) S-Ag has been determined in our laboratory (Fig 1). Comparison of the amino acid sequences indicates a high degree of sequence homology among these species. The molecular weight of S-Ag is approximately 45 kilodaltons and the number of residues are 403, 404 or 405 as shown in Figure 1. Some regions of internal similarity were detected, perhaps reflecting a repeated structure of S-Ag. Searching the data bank revealed no extensive sequence homology between S-Ag and other proteins. However, there were local regions of sequence similarity with α-transducin (Wistow et al., 1986. Shinohara et al., 1988a).

Analysis of nucleotide sequences of S-Ag cDNA also showed a high degree of sequence homology among the mouse, rat, bovine and human (Shinohara et al., 1988a. Yamaki et al., 1987. Yamaki et al., 1988. Tsuda et al., 1989. Abe et al., 1989).

```
        1       10        20        30        40        50
Bo.     MKANKPAPNHVIFKKISRDKSVTIYLGKRDYIDHVERVEPVDGVVLVDPE
Hu. MAASG:AS:SEPN::::::I::::::::I::N:::I:::SQ:Q::::::::::D
Mo. MAACG:TN:S---::::::V::::::::Y::K:::V:::SQ:E::::::::::E
Ra. MAACV:TN:S---::::::V::::::::Y::K:::I:::SQ:E::::::::::E
                60        70        80        90        100
        LVKGKRVYVSLTCAFRYGQEDIDVMGLSFRRDLYSFQVQVFPPVGASGAT
        :::::K:::T:::::::::-:V:WI::T::::::::R:::Y:::::ASTP
        :::::K:::T:::::::::E:I:VM::T::::::::R:::Y:::::MSVL
        :::::K:::T:::::::::E:I:VI::T::::::::R:::Y:::::MSAP
                110       120       130       140       150
        TRLQESLIKKLGANTYPFLLTFPDYLPCSVMLQPAPQDVGKSCGVDFEIK
        :K::E::L::::S:::::::::::::::::::::::::S:::::::::V:
        :Q::E::L::::D:::::::::::::::::::::::::V:::::::::V:
        :Q::L::L::::D:::::::::::::::::::::::::V:::::::::V:
                160       170       180       190       200
        AFATHSTDVEEDKIPKKSSVRLLIRKVQHAPRDMGPQPRAEASWQFFMSD
        :::TDS::A::::::::::::S:::S::RW:LEL::::R::::::::::E
        :::SDI::P::::::::::::L:::K::HA:PEM::::S::::::::::D
        :::TDI::A::::::::::::L:::K::HA:PEM::::C::::::::::D
                210       220       230       240       250
        KPLRLAVSLSKEIYYHGEPIPVTVAVTNSTEKTVKKIKVLVEQVTNVVLY
        :::H:A:C:TR:TNF:::::::::D:::N:E:T:::::ASW::VA:::::
        :::N:S:S:SK:IYF:::::::::T:::N:D:V:::::VSV::IA:::::
        :::H:S:S:SK:IYF:::::::::T:::N:E:V:::::VSV::IA:::::
                260       270       280       290       300
        SSDYYIKTVAAEEAQEKVPPNSSLTKTLTLVPLLANNRERRGIALDGKIK
        :::::V:P::M::A::::P:::T:::::T:L:::::::::::::::::::
        :::::V:P::S::T::::Q:::T:::::V:V:::::::::::::::::::
        :::::V:P::S::T::::Q:::T:::::V:V:::::::::::::::::::
                310       320       330       340       350
        HEDTNLASSTIIKEGIDKTVMGILVSYQIKVKLTVSGLLGELTSSEVATE
        :::::::::::::::::R::L::::::Q:::::::::F::::::::::::
        :::::::::::::::::R::M::::::H:::::::::F::::::::::::
        :::::::::::::::::R::M::::::H:::::::::F::::::::::::
                360       370       380       390       400
        VPFRLMHPQPEDPDTAKESFQDENFVFEEFARQNLKDAGEYKEEKTDQEA
        :::::::::::::--::::Y::A:L:::::::H::::A::AE:G:R:KND
        :::::::::::::--::::V::E:L:::::::Q::::T::NT:G:K:EDA
        :::::::::::::--::::V::E:L:::::::Q::::T::NT:G:K:ED

         AMDE (404)
         A-:: (405)
         GQ:: (403)
         GQ:: (403)
```

Comparison of the amino acid sequences of rat pineal gland and mouse retinal S-Ag indicated a high degree of homology. Analysis of nucleotide sequences of the above two cDNAs isolated from retinal and pineal gland libraries showed high sequence similarity (Abe et al., 1989). These results suggest that the S-Ag in the retina and pineal gland are virtually identical at the transcript and polypeptide level.

Secondary structure prediction and circular dichroic spectroscopy show that S-Ag is composed predominantly of a β-sheet conformation (Shinohara et al., 1987).

5. IMMUNOGENIC SITES IN S-AG

A) T-CELLS

The knowledge of S-Ag amino acid sequences allowed us to identify the uveitopathogenic sites.

Figure 1. Comparison of the amino acid sequences of four different species of S-Ag. Alignment of the amino acid sequences was achieved by the method of Kyte and Doolittle (1982). Colons (:) indicate the identical amino acid residues. Gaps (-) have been introduced in the sequences where it was essential for alignment purpose. Numbers above the sequence are based upon the bovine S-Ag sequence. Numbers in parentheses at the end of the sequences indicate the total amino acid residues in S-Ag. Bo = bovine S-Ag from retina. Hu = human S-Ag from retina. Mo = mouse S-Ag from retina. Ra = rat S-Ag from pineal gland. The one-letter code has been used in order to conserve space (A = alanine: R = arginine: N = asparagine: D = aspartic acid: G = glycine: H = histidine: C = cysteine: Q = glutamine: E = glutamic acid: I = isoleucine: L = Leucine: K = lysine: M = methionine: F = phenylalanine: P = proline: S = serine: T = threonine: W = tryptophan: Y = tyrosine: V = valine).

Initially, a series of 23 oligopeptides were synthe- sized corresponding to the entire amino acid seque- nce of the bovine S-antigen, and tested for their ability to induce EAU in Lewis rats and guinea pigs (Fig.1) (Donoso et al., 1986. Donoso et al., 1987a, 1987b. Singh et al., 1988a. Singh et al ., 1988b. Singh et al., 1989a). In these studies, four synthetic peptides, designated peptide M (18 amino acid residues) and peptide N (22 amino acid resid- ues) corresponding to amino acid position 282-320 in bovine S-Ag consistently induced EAU when immunized at a 50-100 ug dose (major sites). In addition, other synthetic peptides, peptide K (20 amino acid residues) and peptide 3 (20 amino acid residues), corresponding to amino acid position 221-260 also induced mild EAU with higher doses (100-500ug) of immunization (minor sites) (Donoso et al., 1987a. Singh et al., 1989a). In general, the various T cell epitopes from a single protein are not recognized with equal efficiency and there appears to be a hierarchy in the ability of any of these epitopes to be recognized by T cells (see Gammon et al., 1987). The primary and secondary structure of these peptides may contribute to the immunodominance of certain peptides.

Consistent with the high degree of sequence homology of S-Ag, the sequence of uveitopathogenic peptides M and N are virtually identical in all species tested (Table 2) (see Shinohara et al., 1988a). Peptides M and N are also identical in the S-Ag of retina and pineal gland (Abe et al., 1989). It is reasonable that these conserved peptides (M and N) induce major EAU and EAP in various animals. The peptides K and 3 have different sequences among the species shown in Table 2. The EAU induced by peptide K and 3 of bovine S-Ag always shows mild disease in Lewis rats (Singh et al., 1988a. Donoso et al., 1987a). The mild form of EAU may be explained by the differences in antigen sequence.

The minimal length of the peptide which can be recognized by T cells is generally 5-8 residues. The length of the uveitopathogenic determinant of peptide M and N is currently 12 amino acid residues (Donoso et al., 1987a). A recent study in our laboratory, however, indicates that the minimal length of the determinants may be less than 12 residues (In preparation).

High sequence similarity in the hydrophobicity profiles were reported between the peptides M and K and peptides N and 3 (Shinohara et al., 1988a. Tsuda et al., 1988. Yamaki et al., 1988). however, the T lymphocytes of rats immunized with one of the peptides did not cross-react with other peptides

TABLE 2 Uveitopathogenic sites in bovine S-Ag.

	Major Uveitopathogenic Sites	
	M	N
Bo	DTNLASSTIIKEGIDKIT	VPLLANNRERRGIALDGKIKHE
Hu	:::::::::::::::R::	::::::::::::::::::::::
Mo	:::::::::::::::R::	::::::::::::::::::::::
Ra	:::::::::::::::R::	::::::::::::::::::::::
	Minor Uveitopathogenic Sites	
	3	K
Bo	PVTVAVTNSTEKTVKKIKVL	VEQVTNVVLYSSDYYIKTVA
Hu	::::D:::N:E:T:::::AS	W::VA::::::::::V:P::
Mo	::::T:::N:D:V:::::VS	V::IA::::::::::V:P::
Ra	::::T:::N:E:V:::::VS	V::IA::::::::::V:P::

Amino acid sequence similarities of pathogenic sites in bovine (Bo), human (Hu), mouse (Mo) and rat (Ra) S-Ag. Identical residues are indicated by a colon (:). The range of bovine pathogenic peptides N (282-302), M (303-317), K (241-260) and 3 (221-239) are indicated by lines. Peptides N and M are "major" and peptide 3 and K are "minor" uveitopa- thogenic sites. The one-letter code is used to designate amino acid residues.

(Singh et al., 1989a). These results indicate that peptides M,N,K and 3 in bovine S-Ag are independent epitopes.

Although not true in every case, two types of immunodominant oligopeptides are predicted by Rothbard and Taylor (1988) when whole molecules are used for priming. One type of immunodominant peptide is an amphipathic α-helix conformation (DeLisi and Berzofsky, 1985. Margalit et al., 1987) and other type is the peptide which contain the motif of glycine or charged amino acid followed by 2-3 hydrophobic residues and then a hydrophilic amino acid. The peptides M, K and 3 have neither amphipathic α-helical conformation (Shinohara et al., 1987)) nor the latter case of predicted immunodominant sequence. Peptide N, however, does have predict immunodominant sequence, i.e, it has glycine followed by isoleucine-alanine-leucine residues (hydrophobic residues) and then aspartic acid (charged residue).

B) B-CELLS

In contrast to T-cells, the determinant sites (epitopes) of B-cells against S-Ag are somewhat different. The epitopes of B-cells have been characterized using several mouse monoclonal antibodies (Donoso et al., 1986) and polyclonal antibodies from rats, rabbits and guinea pigs (Knospe et al., 1989) derived from native bovine S-Ag. The strongest epitopes are found at one or two sites near the carboxyl end of S-Ag. One monoclonal antibody (MAbA9-C6) is bound to peptide 3 (one of the minor uveitopathogenic sites). This binding and the fact that MAbA9-C6 delays or prevents EAU suggest that this region of peptide may be important in the pathogenesis of EAU (Donoso et al., 1986). Alternatively B cells may play some role in the pathogenicity of EAU.

6. MOLECULAR MIMICRY IN EAU

A) PEPTIDES

Knowledge of the uveitopathogenic sites of S-Ag enabled us to search for similar sequences in foreign antigens. We compared amino acid sequences in a variety of proteins listed in the National Biomedical Research Foundation data base for sequence homology to peptide M and found sequence similarity to a variety of proteins including viral, fungus, bacteria and others as shown in Table 3 (Donoso et al., 1987a. Singh et al., 1989b, 1989c, 1989d). We did not find any protein having sequence homology with peptide N. The best match with peptide M was of Baker's yeast (*Saccharomyces cerevisiae*) histone H3 (Brandt , van Holt 1982. Singh et al., 1989c) and *E. coli* hypothetical protein

TABLE 3 Nonself proteins with sequence similarity to peptide M of S-Ag.

Protein designation	Amino acid sequence
Peptide M.	D T N L A S S T I I K E
Yeast histone H3	D T N L A A I H A K R V
E.coli elongation factor	A R H L A A S I A F K E
E.coli hypothetical protein	L A N L A S S T Q L C K
Potato inhibitor IIa	D T N I A S Y K S U C E
Hepatitis B virus protein	L T N L L S S N L S W L
Baboon endogenous virus	P T N L A K V R T I T Q
Moloney murine leukemia v.	P T N L A K V K G I T Q
Moloney murine sarcoma v.	P T N L A K V K G I T Q
AKV murine leukemia virus	P T N L A K V K G I T Q

The similar amino acid sequences were extracted from the National Biomedical Research Foundation data base. Underlines indicate identical amino acid residues with peptide M. The one-letter code is used to designate amino acid residues.

(Post et al., 1980. Cerretti et al., 1983. Singh et al., 1989b), where the six or five amino acids are identical. Potato proteinase inhibitor and proteins in E. coli elongation factor also have a four or five consecutive amino acids sequence homology. Among viral protein we found several homologous proteins as shown in Table 3.

Table 4. Uveitopathogenicity of various doses of yeast histone H3 and E. coli hypothetical protein in Lewis rats

Immunizing Agent	Immunizing Dose, µg	No.with EAU/ No.immunized
Histone H3	100	2/4
Histone H3	200	3/4
Histone H3	400	15/15
Native Histone H3	500	6/6
E. coli ECHP 11	400	2/5
E. coli ECHP 11	500	3/5
E. coli ECHP 11	2000	5/5
E. coli ECHP 9	400	2/5
E. coli ECHP 9	500	2/5
E. coli ECHP 9	2000	5/5
Peptide M	50	5/5
Bovine S-Antigen	100	4/5
PBS	-	0

Groups of four to six female Lewis rats weighing 150 to 200 g were immunized in one hind footpad with these peptides and CFA. After one week, the eyes were examined daily by slit lamp biomicroscope for the development of EAU. Amino acid sequence of immunizing agents histone H3 is DTNLAAIHAKRVTIQ- K. Native histone H3 was purified from Baker's yea- st (see text). E. coli ECHP 11 and E. coli ECHP 9 are E.coli hypothetical protein and peptide sequen- ces are LANLASSTQLCKKN and DWLANLASSTQLCK respectively. The sequence of peptide M is shown in table 1.

Oligopeptides of histone H3, two E. coli proteins, potato inhibitor IIa, and five viral proteins in Table 3 were synthesized by conventional solid phase chemistries on resin using an automated peptide synthesizer (SAM II: Biosearch, Inc., San Rafel, CA) and purified by gel filtration , column chromatography and reverse phase HPLC (HPLC: Vydac 218 Tp54 column: Vydac, Hesperia, CA).

The incidence of EAU after immunization of Lewis rats with different doses of histone H3, E. coli hypothetical protein (table 4), potato inhibitor IIa, and five viral peptides and peptide M are summarized in the table 6 (see chapter 10). At 100 and 200 μg of histone H3, 50% of the animals developed disease while at the 400 μg dose, all animals developed disease by day 21 (Singh et al., 1989c). E. coli hypothetical peptide and other peptides also induced EAU at higher doses (500-2,000 μg) (Singh et al., 1989b).

The first clinical signs of an inflammatory response were hyperemia of the conjunctivae, iris vasodilation and the appearance of exudate and cells in the anterior chamber and vitreous cavity. Four to six days following the onset of EAU, the inflammatory activity gradually subsided, the cells in the anterior chamber disappeared, and the inflammed eyes became quiet clinically (Singh et al., 1989b). Histopathologically, a more severe inflammatory response was present in the rats immunized with the higher doses of the peptide (400 μg or more) as compared to the lower doses (100-200 μg) where the disease was found to be mild. In the severely inflamed eyes, the retinas were totally detached, showing a complete loss of the rod outer segments of the photoreceptor cells. A subretinal exudate was present containing mononuclear lymphocytes.

In addition, rats with EAU showed an associated EAP characterized by a lymphocytic infiltration of the subcapsular area of the pineal glands (Singh et al., 1989a, 1989b, 1989c). Other peptides from E.

coli protein, potato inhibitor and four viral proteins also induced EAU. In these cases, the histology of the diseased retina and pineal gland was indistinguishable from EAU induced by peptide M (Singh et al., 1989b).

B) NATIVE PROTEIN

To confirm our results using the synthetic polypeptides, we purified native histone, which is highly conserved among different species, from cultures of S. cerevisiae and immunized a group of five Lewis rats with the native protein. Our preparation contained highly purified histones (more than 95%) although histone H3 comprized only 10 to 15% of the preparation. Five rats were immunized and all rats developed EAU 20-24 days postimmunization with 500 µg of native histone and CFA (Singh et al., 1989c). There was cellular infiltration in the anterior chamber as well as in the vitreous cavity and posterior iris synechiae. Histological examination of the eye showed a severe inflammatory response with destruction of the photoreceptor cell layer of the retina, similar to that induced by native S-antigen, peptide M or histone H3. Rats with EAU showed an associated pinealitis characterized by lymphocytic infiltration of the pineal gland (Singh et al., 1989c). No histopathological change was observed in eyes or pineal glands of control animals injected with PBS and CFA emulsion. We also injected 18 other oligopeptides of S-antigen into Lewis rats. None of rats developed EAU (Donoso et al., 1987a. Singh et al., 1989c). Rats injected with a bovine thymus histone mixture (Sigma Chemical Co. St Louis, MO) with CFA did not develop any clinical sign of EAU for five weeks after immunization. These results illustrate that native histone as well as the synthetic preparation are effective agents for the induction of EAU. However, we could not induce EAU with 500 µg of highly purified native potato proteinase inhibitor (a kind gift from Dr. Teruo Iwasaki. Kobe University, Japan).

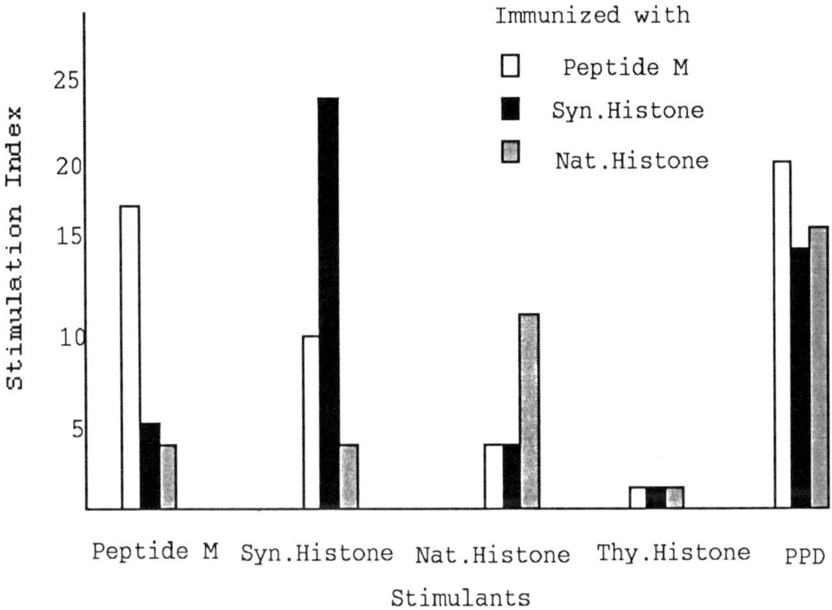

Figure 2 Proliferative response of primed lymph node cells from rats immunized with different antigens. Animals were immunized with 200 μg of peptide M, histone H3 peptide, or native histone from S. Cerevisiae. The concentrations of peptide M, histone H3 peptide and native histone were 2.0 μg/ml for immunogen and 20 μg/ml for cross-reactive antigen. The concentrations of calf thymus histone, PPD, were 20, 20 and 2μg/ml, respectively. Stimulation index value was calculated as mean cpm of test cultures divided by mean cpm of the unstimulated cultures. S.D.of the mean cpm of triplicates was routinely < 20%. The assay was repeated three times with similar result. T cells from animals immunized with other fragments of S-Ag did not show significant proliferation (stimulation index <2) when stimulated with peptide M or histone H3 peptide.

8. LYMPHOCYTE CROSS-REACTION

In order to comfirm the lymphocyte cross-reaction, responsiveness of lymph node cells from rats immunized with peptide M, histone H3 and native histone were tested (Fig. 2). Lymph node cells from rats immunized with antigens reacted well against immunizing antigen. In addition, lymph node cells from rats immunized with peptide M also responded well to histone H3 or native histone (Singh et al., 1989c. Singh et al., 1989d). Similarly, cells from rats immunized with either histone H3 or native histone showed significant proliferation when cultured with peptide M. None reacted with calf thymus histone. There was a significant response to Purified Protein Derivative (PPD, a component of CFA) and Con A (Singh et al., 1989c). These results indicate that peptide M and histon H3 peptide are similar enough to cross-react.

9. ADOPTIVE TRANSFER

Lymph node cells of animals immunized against peptide M and histone H3 stimulated with either peptide M or histone H3 induced EAU between 6-12 days when injected intraperitoneally in naive syngeneic rats (Table 5). Histopathology showed severe inflammation in the uveal tract and a complete loss of photoreceptor cells in EAU-positive animals which was indistinguishable from rats induced by peptide M, histone H3 or E. coli hypothetical protein or S-antigen (Singh et al., 1989d).
Thus, EAU is adoptively transferred to the naive rats by lymph-node cells stimulated with either peptide M or histone H3.

10. MOLECULAR MIMICRY IN OTHER SYSTEM

A) T-CELLS

EAU and other experimental autoimmuno inflam-

Table 5 Adoptive transfer of EAU by intraperitoneal injection of sensitized lymph node cells.

Donor Sensitization	Stimulant in Culture Peptide M	Histone H3	None
Peptide M	+	+	-
Histone H3	+	+	-
Saline	-	-	-

Groups of five rats were immunized with peptide M, histone H3 peptide or PBS. Animals were killed at the onset of EAU and cultures of cervical lymph node cells were set up in tissue culture flasks as described (Singh et al., 1989d) using peptide M or histone H3 peptide (20 μg/ml) as stimulant. Cultures were incubated for 3 days and, after incubation, cells were harvested, separated on a percoll gradient, washed twice, and injected ($50x10^6$ cells/rat) into naive syngeneic rats. The rats were observed daily after 3 days by slit lamp biomicroscope for the development of EAU. The experiment was repeated three times and similar result was observed. + indicates 3-4 out of 5 rats developed EAU. - indicates 0 out of 5 rats developed EAU.

mations are predominantly induced by T-cells as mantioned above. The evidence in support of the role of molecular mimicry in the pathogenesis of autoimmune diseases by T cells is summarized in Table 6. These studies are based on amino acid sequence similarities between foreign proteins and peptide M and induction of EAU with these peptides. In experimental allergic encephalomyelitis (EAE), an animal model of multiple sclerosis caused by immunization with myelin basic protein (MBP), Fujinami and Oldstone showed (1985) that a six amino acid sequence of hepatitis B virus polymerase (HBVP) was identical to the encephalitogenic site of MBP. Rabbits injected with HBVP synthetic peptide showed antibody responses that cross-reacted with MBP. Peripheral blood mononuclear cells proliferated when incubated with either MBP or HBVP. This finding is the first example of molecular mimicry reported in which EAE is induced by T cell-mediated autoimmune inflammation.

TABLE 6 Nonself protein or peptides which induced experimental autoimmune inflammation in Lewis rats

Nonself protein/ Self protein	Experimental autoimmune inflammation	Ref.
Hepatitis virus/ Myelin basic protein	EAE	1#
Baker's Yeast histone H3/ S-Ag	EAU	2
Baker's Y. native histone H3*/ S-Ag	EAU	2
E.coli hypothetical protein/ S-Ag	EAU	3
Potato inhibitor/ S-Ag	EAU	4
Hepatitis virus/ S-Ag	EAU	4
Baboon virus/ S-Ag	EAU	4
Moloney murine sarcoma virus/ S-Ag	EAU	4
Moloney murine leukemia virus/ S-Ag	EAU	4
AKR murine leukemia virus/ S-Ag	EAU	4

* indicates native protein. The rest of them indicate synthetic peptides. References 1#: Fujinami and Oldstone, 1985. 2: Singh et al., 1989c. 3: Singh et al., 1989b. 4: Singh et al., in preparation.

In our report, we have shown that lymphocytes, obtained from rats immunized with either histone H3, peptide M or native histone proteins proliferated in the presence of these antigen and induced EAU in Lewis rats. In addition, potato inhibitor or E. coli hypothetical protein and five viral peptides also induced EAU in Lewis rats as shown in table 6 and lymphocytes immunized with peptide M proliferated well in the presence of all peptides listed in table 6. The pathological and histological observations of EAU induced by these peptides were indistinguishable from the EAU induced by S-Ag or peptide M (Singh et al., 1989b, 1989d).

B) B-CELLS

In contrast to T-cell-mediated molecular mimicry, B-cell (antibody response) mediated cross-reaction has been well described. Srinivasappa et al. reported that 4% of monoclonal antibodies from over 600 antibodies raised against a large panel of DNA and RNA viruses reacted with host proteins (Srinivasappa et al., 1986).

Recently Carnegie and Weise reported a six amino acid sequence homology between MBP and the retrovirus Visna pol, a virus associates with demyelinating disease in sheep (Carnegie and Weise, 1987). Since retroviruses are being investigated for involvement in multiple sclerosis, this observation might inspire a more specific search for a Lentivirus with an ability to mimic MBP sequence in human oligodendrocytes. There is strong evidence that molecular mimicry plays a role in the pathogenesis of celiac disease and ankylosing spondylitis (Oldstone, 1987). Celiac disease, an intestinal malabsorption syndrome, is activated by the wheat protein α-gliadin. α-Gliadin, in turn shares a sequence homology with the region E1b protein of human adenovirus serotype 12 (Ad 12), an adenovirus isolated from the human intestinal tract. Antibody reactive with the Ad 12 E1b protein cross-reacts with α-gliadin (Kagnoff et al., 1984). In another study, 89% of patients with celiac disease showed evidence of previous Ad 12 infection, but no evidence of any infection with other closely related viruses, eg.adenovirus 18 or another enteric virus, echovirus (Kagnoff et al., 1987). These observations suggest that a viral protein may play a role in the pathogenesis of celiac disease perhaps by virtue of immunological cross-reactivity between antigenic determinants shared by viral protein and α-gliadin. However, it is not clear whether viral infection is a primary event or immune activation against α-gliadin is a initial event.

The relationships between HLA-B27 antigen and several microbial agents such as Klebsiella pneumoniae, Yersinia pseudotuberculosis, and shigella has also been investigated (Oldstone 1987). Six consecutive amino acids are identical between the hypervariable domain of HLA-B27.1 and K. pneumoniae nitrogenase. Sera from a significant proportion of HLA-B27 individuals with ankylosing spondylitis (53%) or Reiter's syndrome (29%), but not from appropriate controls, reacted with a synthetic peptide containing the homologous region of HLA-B27.1 (Schwimmbeck et al., 1987). As expected, sera from ankylosing spondylitis and Reiter's syndrome patients also reacted to a peptide derived from the K. pneumoniae nitrogenase. Chen and associates showed that antibody to the hypervariable region of HLA-B27.1 reacted to two strains of Y. pseudotuberculosis, another microbe implicated in ankylosing spondylitis and Reiter's syndrome (Chen et al., 1987). These observations suggest that ankylosing spondylitis and Reiter's syndrome are autoimmune diseases directed against HLA-B27. Induction could be caused by microbes encoding a protein with sequences homology to HLA-B27, and the manifestation of the disease would be related to unusual concentration or presentations of HLA antigen in certain tissues.

11. CONCLUSION AND REMARKS

Studies by us and others provide important clues in understanding autoimmune diseases of the eye and pineal gland. Autoimmune response provoked by molecular mimicry occurs when the nonself and self determinants are similar enough to cross-react, yet different enough to break immunological tolerance (Oldstone, 1987).

Our findings that selected bacterial, fungal, viral and other proteins with sequence similarity to retinal S-Ag can induce EAU added to the evidence supporting molecular mimicry as an etiology of autoimmune uveitis. Synthetic oligopeptides offer a convenient and irreplaceable tool for studying molecular mimicry. They also allow us to obtain valuable information for understanding the underlying mechanisms of autoimmune inflammatory diseases. Finding that a native histone H3 protein induces EAU in Lewis rats is to the best of our knowledge, the first clear examlpe of molecular mimicry at the experimental level in any of systems and should be helpful for understanding the mechamism of lymphocyte processing in the immune system.

Our results indicate that microbe infection may initiate immune system events. In addition to infectious microorganism proteins, common food proteins from yest and potato are also possible initiators of autoimmune inflammation.

Molecular mimicry could very well be a general phenomena in nature and more examples of such mimicry may be found in the near future. The rapid advances in the molecular cloning of cDNAs and genes and the determination of amino acid sequences of the encoded proteins constantly expand sequence information. Sequence information from data banks in conjunction with the use of synthetic oligopeptides allow for rapid progress in finding more examples of molecular mimicry.

Although, genetic predisposition, multiple lymphatic factors and lymphocyte interactions may play major roles in the development of autoimmune diseases, molecular mimicry between self and nonself proteins may be a underlying primary initiator of autoimmune inflammation. Studies of cell-mediated immune responses and of antibodies in animal models and patients with uveitis should help test further the role of molecular mimicry in the pathogenesis of autoimmune inflammatory uveitis as well of autoimmune diseases in general.

REFERENCES

Abe T, Yamaki K, Tsuda M, Singh VK, Suzuki S, McKinnon R, Klein DC, Donoso LA, Shinohara T (1989). Rat pineal S-antigen: sequence analysis reveals presence of α-transducin homologous sequence. FEBS Lett. 247:307-311.

Brandt WF, von Holt B (1982). The primary structure of yeast histone H3. Eur.J.Biochem. 121:501-510.

Bottazzo GF, Todd I, Mirakian R, Belfiore A Pujol-Borrell R (1986). Organ-specific autoimmunity: A 1986 Overview. Immunol. Rev. 94:137-170.

Carnegie PR, Weise MJ (1987). Visna and myelin basic protein. Nature 329:294.

Chan C-C, Mochizuki M, Nussenblatt RB, Palestine AG, McAllister C, Gery I, BenEzra D (1985). T lyphocyte subsets in experimental autoimmune uveitis. Clin. Immunol. Immunopathol. 35:103-110.

Caspi RR, Roberge FG, McAllister CG, M. El-Saied, Kuwabara T, Gery I, Hanna H, Nussenblatt RB (1986). T cell lines mediating experimental autoimmune uveoretinitis (EAU) in the rat. J. Immunol. 136:928-933.

Cerretti DP, Dean D, Davis GR, Bedwell DM, Nomura M (1983). The spc ribosomal protein operon of Escherichia coli: sequence and cotranscription of the ribosomal protein genes and a protein export gene. Nuc. Acids Res. 11:2599-2616.

Chen JH, Kono DH, Yong Z, Park MS, Oldstone MBA, Yu DTY (1987). A Yersinia pseudotuberculosis protein which cross-reacts with HLA-B27. J. Immunol. 139:3003-3011.

Cohen IR (1986). Regulation of autoimmune disease, physiological and therapeutic. Immunol. Rev. 94:5-21.

Cohen IR (1988). The self, the world and autoimmunity. Scientific American, April; PP. 52-60.

de Kozak Y, Sakai J, Thillaye B, Faure J-P (1981). S antigen-induced experimental autoimmune uveo-retinitis in rats. Curr. Eye Res. 1:327-337.
de Kozak Y, Audibert F, Thillaye B, Chedid L, Faure J-P (1979). Effects of mycobacterial hydrosoluble adjuvants on the induction and prevention of experimental autoimmune uveo-retinitis in guinea pigs. Ann. Immunol. (Paris). 130:29-32.
de Kozak Y, Yuan WS, Bogossian M, Faure J-P (1976) Humoral and cellular immunity to retinal antigens in guinea pigs. Mod. Probl. Ophthalmol. 16:51-58.
DeLisi C, Berzofsky JA (1985) T-cell antigenic sites tend to be amphipathic structure. Proc. Natl. Acad. Sci. USA. 82:7042-7052.
Donoso LA, Merryman CF, Shinohara T, Dietzschold B, Wistow G, Craft C, Morley W, Henry R (1986). S-antigen: identification of the MAbA9-C6 monoclonal antibody binding site and the uveitopathogenic sites. Curr. Eye Res. 5:995-1004.
Donoso LA, Merryman CF, Sery T, Shinohara T, Dietzschold B, Smith A, Kalsow CM (1987a). S-Antigen: Characterization of a pathogenic epitope which mediates experimental autoimmune uveitis and pinealitis in Lewis rats. Curr. Eye Res. 6:1151-1159.
Donoso LA, Merryman CF, Shinohara Y, Sery TW, Smith A (1987b). S-Antigen: experimental autoimmune uveitis following immunization with a small synthetic peptide. Arch. Ophthalmol. 105:838 840.
Dory C, Faure J-P (1977). Isoelment et characterisation d'un antigene retinien de la retine induisant l'uveo-retinite autoimmune experimentale. Ann. Immunol.(Inst. Pasteur) 128C:229-232.
Dyrberg T, Oldstone MBA (1986). Peptides as probes to study molecular mimicry and virus-induced autoimmunity. Cur. Top. Microbiol. Immunol. 130:25-37.
Faure J-P (1980). Autoimmunity and the retina. Curr. Top. Eye Res. 2:215-302.
Fujinami RS, Oldstone BA (1985). Amino acid homology between the encephalitogenic site of myelin basic protein and virus: mechanism for autoimmunity. Science 230:1043-1045.

Gammon G, Shastri N, Cogswell J, Wilbur S, Sadegh-Nasseri S, Krzych U, Miller A, Sercaze E (1987). The choice of T-cell epitopes utilized on a protein antigen depend on multiple factors distant from, as well as at the determinant site. Immunol. Rev. 98:53-73.

Gery I, Mochizuki M, Nussenblatt RB (1986). Retinal specific antigens and immunopathogenic processes they provoke. *in* Progress in retinal research. vol.5 (Osborne N, and Chader J, eds) Pergamon press, New York, pp.75-109.

Hirose S, Singh VK, Donoso LA, Shinohara T, Kokate S, Tanaka T, Kuwabara T, Yamaki K, Gery I, Nussenblatt RB (1989). An 18mer peptide derived from the retinal S-antigen induces uveitis and pinelitis in primates. Clin. Exp. Immunol. 627:627-632.

Kagnoff MF, Austin RK, Hubert JJ, Bernardin JE, Kasarda DD (1984). Possible role for a human adenovirus in the pathogenesis of celiac disease. J. Exp. Med. 160:1544-1557.

Kagnoff MF, Paterson YJ, Kumar PJ, Kasarda DD, Carbone FR, Unsworth DJ, Austin RK (1987). Evidence for the role of a human intestinal adenovirus in the pathogenesis of coeliac disease. Gut 28:995-1001.

Kakimoto K, Katsuki M, Hirofuji T, Iwata H, Koga T, (1988) Isolation of T-cell line capable of protecting mice against collegen-induced arthritis. J. Immiunol. 140:78-83.

Kalsow CM, Wacker WB (1977). Pineal reactivity of anti-retina sera. Invest. Ophthalmol. Vis. Sci. 16:181-184.

Kalsow CM, Wacker WB (1978). Pineal gland involvement in retina induced experimental alleric uveitis. Invest. Ophthalmol. Vis. Sci. 17:774-783.

Knospe V, Gregerson DS, Donoso LA, Rathbun WB (1989). Mapping the immunogenic region of S-Ag: Influence of MHC, IGH and species differences is small.Invest.Ophthalmol.Vis.Sci.30:(Suppl) 81.

Kong YM, Bagnasco M, Canonica GW (1986). How do T cells mediate autoimmune thyroiditis. Immunol. Today 7: 337-339.

Kumar V, Kono DH, Urban JL, Hood L (1989). The T-cell receptor repertoire and autoimmune diseases. Ann. Rev. Immunol. 7:657-682.
Kuhn H, Hall SW, Wilden U (1984). Light-induced binding of 48-KDa protein to photoreceptor membranes is highly enhanced by hphosphorylation of rhodopsin. FEBS Lett. 176:473-478.
Kyte J, Doolittle RF (1982). A simple method for displaying the hydropathic character of a protein. J. Molec. Biol. 157:105-132.
Linington C, Izumo S, Suzuki M, Uemura K, Meyermann R, Wekerle H, (1984). A permanent rat T-cell line that mediates experimental allergic neuritis in the Lewis rat in vivo. J. Immunol. 133:1946-1950.
Margalit H, Spouge JL, Cornette JL, Cease KB, DiLisi C, Berzofsky JA (1987). Prediction of Immunodominant helper T-cell antigenic sites from the primary sequence. J. Immunol. 138:2213-2219.
Mochizuki M, Kuwabara T, Chan C-C, Nussenblutt RB, Metcalfe DD, Gery I (1984). An association between suscepitibility to experimental autoimmune uveitis and choroidal mast cell numbers. J. Immunol. 133: 1699-1701.
Mochizuki M, Kuwabara T, McAllister C, Nussenblatt RB, Gery I (1985). Adoptive transfer of experimental autoimmune uveoretinitis in rats: immunopathogenic mechanisms and histological features. Invest. Ophthalmol. Vis. Sci. 26:1-9
Nussenblatt RB, Kuwabara T, de Monasterio FM, and Wacker WB (1981). S antigen uveitis in primates. A new model for human disease. Arch. Ophthalmol. 99: 1090-1092.
Nussenblatt RB, Gery I, Ballintine EJ, Wacker WB (1980). Cellular immune responsiveness of uveitis patients. Am. J. Ophthalmol. 89:173-179.
Oldstone MBA (1987). Molecular mimicry and autoimmune disease. Cell 50:819-820.
Oldstone MBA, Notkins AL (1986). Molecular mimicry. In Concepts in Viral Pathogenesis. Vol.II (Notkins AL, and Oldstone MBA, eds). Springer-Verlag, New York, pp. 195-202.

Pfister C, Chabre M, Plouet J, Tuyen v, de Kozak Y, Faure JP, Kuhn H (1985). Retinal S-antigen identified as the 48K protein regulateing light-dependent phosphdiesterase in rod cells. Science 228:891-893.

Post LE, Arfsten AE, Davis GR, Nomura M, (1980). DNA sequence of the promoter region for the a ribosomal protein operon in Escherichia coli. J. Biol. Chem. 255:4653-4659.

Rothbrard JB, Taylor WR (1988). A sequence pattern common to T-cell epitopes. EMBO J. 7:93-100.

Schwartz RH (1985). T-Lymphocyte recognition of antigen in association with gene products of the major histocompatibility complex. Ann. Rev. Immunol. 3: 237-261.

Schwimmbeck PL, Yu DTY, Oldstone MBA (1987). Autoantibodies to HLA B27 in the sera of HLA B27 patients with ankylosing spondylitis and Reiter's syndrome. J. Exp. Med. 166:173-181.

Shinohara T, Dietzschold B, Craft CM, Wistow G, Early J, Donoso LA, Horwitz J, Tao R (1987). Primary and secondary structure of bovine retinal S-antigen (48-kDa protein). Proc. Natl. Acad. Sci. USA 84:6975-6579.

Shinohara T, Donoso LA, Tsuda M, Yamaki K, Singh VK (1988a). S-antigen: Structure, function and experimental autoimmune uveitis (EAU). *In* Progress in Retinal Research vol.8. (Osborne N and Chader J, eds). Pergamon Press, New York. pp. 51-66,

Shinohara T, Singh VK, Hirose K, Yamaki K, Donoso LA, Kuwabara T, Gery I, Nussenblatt RB (1988b). S-Antigen autoimmune uveitis induced in Lewis rats and monkey following immunization with synthetic peptides. Invest. Ophthalmol. Vis. Sci. 27:(Suppl) 372.

Shoenfeld Y, Isenberg DA (1988). Mycobacteria and autoimmunity. Immunol. Today 9: 178-181.

Singh VK, Nussenblatt RB, Donoso LA, Yamaki K, Chen C, Shinohara T (1988a). Identification of a uveitopathogenic and lymphocyte proliferation site in bovine S-antigen. Cell. Immunol. 115:413-419.

Singh VK, Yamaki K, Donoso LA, Shinohara T (1988b). S-Antigen: Experimental antoimmune uveitis induced in guinea pigs with two synthetic peptides. Curr. Eye Res. 1:87-92.

Singh VK, Donoso LA, Yamaki K, Shinohara T (1989a). Uveitopathogenic sites in bovine S-antigen. Autoimmunity 3:177-187.

Singh VK, Yamaki K, Abe T, Shinohara T (1989b). Molecular mimicry between uveitopathogenic sites of retinal S-antigen and Escherichia coli protein: induction of experimental autoimmune uveitis and lymphocyte cross-reaction. Cell. Immunol. in press.

Singh VK, Yamaki K, Donoso LA, Shinohara T (1989c). Molecular Mimicry: Yeast histone H3 induced experimental autoimmune uveitis. J. Immunol. 142: 1512-1517.

Singh VK, Yamaki K, Donoso LA, Shinohara T (1989d). Sequence homology between yeast histone H3 and uveitopathogenic site of S-antigen: Lymphocyte cross-reaction and adaptive transfer of the disease. Cell. Immunol. 119:211-221.

Srinivasappa J, Saegusa J, Prabhakar BS, Gentry MK, Buchmeier MJ, Wiktor TJ, Koprowski H, Oldstone BA, Notkins AL (1986). Molecular mimicry: frequency of reactivity of monoclonal antiviral antibodies with normal tissues. J. Virol.57:397-401.

Tsuda M, Syed M, Bugra J, Whelan JP, McGinnis JF, Shinohara T (1988). Structural analysis of mouse S antigen. Gene 73:11-20.

Wacker WB, Lipton MM (1965). Experimental allergic uveitis 1. Homologous retina as uveitogenic antigen. Nature 206:253-254.

Wacker WB, Donoso LA, Kalsow CM, Yankeelov JA Jr, Organisciak DT (1977). Experimental allergic uveitis. isolation, characterization and localization of a soluble uveitopathogenic antigen from bovine retina. J. Immunol. 119:1949-1958.

Watt TH, Brian A, Kappler JW, Marrack P, McConnell HM (1984). Antigen presentation by supported planar membranes containing affinity-purified I-A^d. Porc. Natl. Acad. Sci. USA. 81:7564-7568.

Wick G, Hala K, Wolf H, Ziemiecki A, Sundick RS, Stoffler-Meilicke M, DeBaets M (1986). The role of genetically determined primary alterations of the target organ in the development of spontaneous autoimmune thyroiditis in obese strain (OS) chickens. Immunol. Rev. 94:113-136.

Wiggert B, Chader G (1985). Monkey interphoto-receptor retinoid-binding protein (IRBP): Interphotoreceptor Matrix in Health and Desease (C.D. Bridges and A.J.Adler, eds) pp,89-110. Alan R. Liss, New York.

Wistow GJ, Katial A, Craft C, Shinohara T (1986). Sequence analysis of bovine retinal S-antigen. Relationships with α-transducin and G-proteins. FEBS Lett. 196:23-28.

Yamaki K, Takahashi M, Sakuragi S, Matsubara K (1987). Molecular cloning of the S-antigen cDNA from bovine retina. Biochem. Biophys. Res. Commun. 142:904-910.

Yamaki K, Tsuda M, Shinohara T (1988). The sequence of human retinal S-antigen reveals similarities with α-transducin. FEBS Lett. 234:39-43.

Molecular Biology of the Retina: Basic and Clinically Relevant Studies, pages 191–219

MOLECULAR BASIS OF ORNITHINE AMINOTRANSFERASE DEFECT IN GYRATE ATROPHY

George Inana, Yoshihiro Hotta, Carmelann Zintz, Carolyn Chambers, Nancy G. Kennaway, Richard G. Weleber, Akira Nakajima and Takashi Shiono

National Eye Institute, NIH, Bethesda, Maryland 20892 (G.I., Y.H., C.Z., C.C.), Departments of Ophthalmology (R.G.W.) and Medical Genetics (N.G.K.), The Oregon Health Sciences University, Portland, Oregon 97201, Department of Ophthalmology, Juntendo University School of Medicine, Tokyo, Japan (A.N.) and Department of Ophthalmology, Tohoku University School of Medicine, Sendai, Japan (T.S.)

Gyrate atrophy (GA) is a rare autosomal recessive degenerative disease of the retina and choroid of the eye that leads to blindness (Valle and Simell, 1983). Simell and Takki demonstrated the hyperornithinemia present in patients with GA in 1973 (Simell and Takki, 1973), and a generalized deficiency in the mitochondrial enzyme, ornithine aminotransferase (OAT), was reported in 1977 (Valle et al., 1977; Trijbels et al., 1977; O'Donnell et al., 1977). The obligate heterozygous carriers have 50% of the normal level of OAT activity in their cells, in agreement with the autosomal recessive pattern of inheritance observed. The deficiency of OAT activity in the fibroblasts of two GA patients was shown by an immunoassay technique to be due to a decreased concentration of the enzyme (Ohura et al., 1984). Despite the generalized nature of the OAT deficiency in GA, the eye appears to be the only site of significant pathology in this disease (Valle and Simell, 1983).

Ornithine aminotransferase is a nuclear-encoded, pyridoxal phosphate-requiring, mitochondrial matrix enzyme that catalyzes the interconversion of ornithine, glutamate, and proline (Peraino and Pitot, 1963). OAT is the main catabolic enzyme of ornithine, and a defect in OAT would result in hyperornithinemia as seen in GA. The enzyme is present in many mammalian tissues and has been purified from various tissue sources, including rat liver (Shiotani et al., 1977), kidney (Sanada et al., 1970), brain

(Deshmukh and Srivastava, 1984), and human liver in a crystalline-pure form (Ohura et al., 1982). The rat liver and kidney OAT has been shown to be synthesized as a 49-kilodalton precursor molecule in the cytoplasm, which is then transported to the mitochondrion where it is processed into the 45-kilodalton mature subunit (Mueckler et al., 1982).

Thus, GA is unique among the hereditary retinal degenerative diseases in that the underlying biochemical defect is known. Although the exact mechanism by which the OAT deficiency and hyperornithinemia lead to the chorioretinal degeneration is not known, this disease still offers a great opportunity to study the molecular basis of the genetic defect in OAT using a molecular biological approach. Since the underlying biochemical defect is known, one can take a relatively straightforward 'forward' molecular genetic approach, i.e., cloning and examining the gene for the defective enzyme, in studying this disease. Elucidation of the molecular genetics of GA may yield information that may be useful for genetic counseling and future gene therapy for this disease.

In order to study GA at the gene level, we had constructed a molecular probe for the human OAT in the form of a cDNA clone (Inana et al., 1986). We utilized the λgt11 expression cloning system developed by Young and Davis for cloning the OAT cDNA (Young and Davis, 1983). We were able to use the λgt11 approach through a collaboration with a Japanese group who had purified the human OAT and provided it to us (Ohura et al., 1982). An anti-human OAT antibody was prepared with the pure enzyme and used to isolate the OAT cDNA clone.

The OAT cDNA consists of 2073 nucleotides and codes for a 439 amino acid OAT precursor protein with a molecular mass of 48,534 dalton (Fig. 1). Analysis of the cDNA-derived OAT sequence revealed a leader sequence, similar to those found in other mitochondrial proteins of cytoplasmic origin, in the precursor protein. The OAT protein sequence was also shown to be 91% homologous to the rat OAT (Mueckler and Pitot, 1985) and 27% similar to the chicken aspartate aminotransferase (Graf-Hausner et al., 1983) with a conservation of the pyridoxine-binding lysine residue.

A genomic Southern analysis using the human OAT cDNA as a probe indicated that the OAT gene is a very complicated gene family consisting of multiple copies of OAT-related gene sequences (Fig.2). The results of a hybridization experiment using specific probes and varying

```
        10        20        30        40        50        60        70        80        90
AATTCCTGTCCTCAGGCGCTGTCAGATCTGTGGTTTTTCTACTTGAAGGACACAATGTTTTCCAAACTAGCACATTTGCAGAGGTTTGCT
                                          ***  **   ***MetPheSerLysLeuAlaHisLeuGlnArgPheAla
                                             """"""    '''''''''''''''''''''''''''''''''''''

       100       110       120       130       140       150       160       170       180
GTACTTAGTCGCGGAGTTCATTCTTCAGTGGCTTCTGCTACATCTGTTGCAACTAAAAAAACAGTCCAAGGCCCTCCAACCTCTGATGAC
ValLeuSerArgGlyValHisSerSerValAlaSerAlaThrSerValAlaThrLysLysThrValGlnGlyProProThrSerAspAsp
''''''''''''''''''''''''''''''''''''''''''''''''''''''''''''''''''''''''''''''><

       190       200       210       220       230       240       250       260       270
ATTTTTGAAAGGGAATATAAGTATGGTGCACACAACTACCATCCTTTACCTGTAGCCCTGGAGAGAGGAAAAGGTATTTACTTATGGGAT
IlePheGluArgGluTyrLysTyrGlyAlaHisAsnTyrHisProLeuProValAlaLeuGluArgGlyLysGlyIleTyrLeuTrpAsp
                        ---------   ---------------------------------

       280       290       300       310       320       330       340       350       360
GTAGAAGGCAGAAAATATTTTGACTTCCTGAGTTCTTACAGTGCTGTCAACCAAGGGCATTGTCACCCCAAGATTGTGAATGCTCTGAAG
ValGluGlyArgLysTyrPheAspPheLeuSerSerTyrSerAlaValAsnGlnGlyHisCysHisProLysIleValAsnAlaLeuLys
                  ------------------------

       370       380       390       400       410       420       430       440       450
AGTCAAGTGGACAAATTGACCTTAACATCTAGAGCTTTCTATAATAACGTACTTGGTGAATATGAGGAGTATATTACTAAACTTTTCAAC
SerGlnValAspLysLeuThrLeuThrSerArgAlaPheTyrAsnAsnValLeuGlyGluTyrGluGluTyrIleThrLysLeuPheAsn

       460       470       480       490       500       510       520       530       540
TACCACAAAGTTCTTCCTATGAATACAGGAGTGGAGGCTGGAGAGACTGCCTGTAAACTAGCTCGTAAGTGGGGCTATACCGTGAAGGGC
TyrHisLysValLeuProMetAsnThrGlyValGluAlaGlyGluThrAlaCysLysLeuAlaArgLysTrpGlyTyrThrValLysGly

       550       560       570       580       590       600       610       620       630
ATTCAGAAATACAAAGCAAAGATTGTTTTTGCAGCTGGGAACTTCTGGGGTAGGACGTTGTCTGCTATCTCCAGTTCCACAGACCCAACC
IleGlnLysTyrLysAlaLysIleValPheAlaAlaGlyAsnPheTrpGlyArgThrLeuSerAlaIleSerSerSerThrAspProThr

       640       650       660       670       680       690       700       710       720
AGTTACGATGGTTTTGGACCATTTATGCCGGGATTCGACATCATTCCCTATAATGATCTGCCCGCACTGGAGCGTGCTCTTCAGGATCCA
SerTyrAspGlyPheGlyProPheMetProGlyPheAspIleIleProTyrAsnAspLeuProAlaLeuGluArgAlaLeuGlnAspPro

       730       740       750       760       770       780       790       800       810
AATGTGGCTGCGTTCATGGTAGAACCAATTCAGGGTGAAGCAGGCGTTGTTGTTCCGGATCCAGGTTACCTAATGGGAGTGCGAGAGCTC
AsnValAlaAlaPheMetValGluProIleGlnGlyGluAlaGlyValValValProAspProGlyTyrLeuMetGlyValArgGluLeu
                  ------------    ---------------------------------------------

       820       830       840       850       860       870       880       890       900
TGCACCAGGCACCAGGTTCTCTTTATTGCTGATGAAATACAGACAGGATTGGCCAGAACTGGTAGATGGCTGGCTGTTGATTATGAAAAT
CysThrArgHisGlnValLeuPheIleAlaAspGluIleGlnThrGlyLeuAlaArgThrGlyArgTrpLeuAlaValAspTyrGluAsn
                                    --------------------         ----------------------

       910       920       930       940       950       960       970       980       990
GTCAGACCTGATATAGTCCTCCTTGGAAAGGCCCTTTCTGGGGGCTTATACCCTGTGTCTGCAGTGCTGTGTGATGATGACATCATGCTG
ValArgProAspIleValLeuLeuGlyLysAlaLeuSerGlyGlyLeuTyrProValSerAlaValLeuCysAspAspAspIleMetLeu
--------------------------   ----------------------------------------   ----------------

      1000      1010      1020      1030      1040      1050      1060      1070      1080
ACCATTAAGCCAGGGGAGCATGGGTCCACATACGGTGGCAATCCACTAGGCTGCCGAGTGGCCATCGCAGCCCTTGAGGTTTTAGAAGAA
ThrIleLysProGlyGluHisGlySerThrTyrGlyGlyAsnProLeuGlyCysArgValAlaIleAlaAlaLeuGluValLeuGluGlu
  ---------------                                             ------------------------------

      1090      1100      1110      1120      1130      1140      1150      1160      1170
GAAAACCTTGCTGAAAATGCAGACAAATTGGGCATTATCTTGAGAAATGAACTCATGAAGCTACCTTCTGATGTTGTAACTGCCGTAAGA
GluAsnLeuAlaGluAsnAlaAspLysLeuGlyIleIleLeuArgAsnGluLeuMetLysLeuProSerAspValValThrAlaValArg
--------------------------

      1180      1190      1200      1210      1220      1230      1240      1250      1260
GGAAAAGGATTATTAAACGCTATTGTCATTAAAGAAACCAAAGATTGGGATGCTTGGAAGGTGTGTCTACGACTTCGAGATAATGGACTT
GlyLysGlyLeuLeuAsnAlaIleValIleLysGluThrLysAspTrpAspAlaTrpLysValCysLeuArgLeuArgAspAsnGlyLeu

      1270      1280      1290      1300      1310      1320      1330      1340      1350
CTGGCCAAGCCAACCCATGGCGACATTATCAGGTTTGCGCCTCCCCTGGTGATCAAGGAGGATGAGCTTCGAGAGTCCATTGAAATTATT
LeuAlaLysProThrHisGlyAspIleIleArgPheAlaProProLeuValIleLysGluAspGluLeuArgGluSerIleGluIleIle

      1360      1370      1380      1390      1400      1410      1420      1430      1440
AACAAGACCATCTTGTCTTTCTGAGGGTAGCCAGCTGTTTTCAGTGGTCCCTGGGAGCCAGCTGGAGACAGGTGGTCCTGTAAAAGCTTT
AsnLysThrIleLeuSerPheTer   Ter
ATTCCTAATGTGGCACATTCCACTCCCATGAGTCTTCAAAAACTTTTTTTTTGAATATATTTTTTTCAGTTGATACATAATAGAACAAC
GTTTATGAACCTGCCGTTTGCTTTGTAACGTAACTAAATAATGTAATGGCATCTATATTCACTTCAAGTGTTTTGATGTGCATGTGTACT
TCCTAAGGTGAAATGCATCTATATACAGACAGCCTCTAAATCAAGTCCTTCAGTATAATTGATATATGTTTTTATAATTTCCTCACTGGT
ATAAGTGTTTCATATTTGAAAAAGTTATCTCTGGGTATTGCATAAAAGGCTTCATCTTATAAAGTGAAATCATTGTTATTGAATTTTAGG
AAGGATTAATGGTTAAGTGTATATAAAATACTAATATTAAGTAAACTTCATATTGGCCAACACCAGGGTTGTATTCTATGGATGTCATTA
TTTTGAATTAAGAATTAGTGTTTAACATTCCTAAATTGTTTTGAGTGCTTGATTATAATTTGTAAAAAATGTTTATTTTCAATACTTCTT
TAAATTTAAAATAAAGCTTATATTTCAAATGTCAAAAAAAAAAAAAAAAAAAAAAAAAAAAAAAAAAAAAAAAAAAAAAAAAAAAAAAAGG
AAT(2073)======
```

Figure 1. cDNA and protein sequence of the human OAT. The cDNA contains 2073 bases, including 55 adenine residues in the 3′ poly(A) tail. An open reading frame exists from position 1 to 1371 followed by 702 bases of 3′ untranslated sequence. (*), Residues forming the consensus-like sequences; ("), a cluster of purine residues just upstream of the putative translational initiation codon (ATG) at position 55; (′), putative leader sequence; (><), putative cleavage site of the leader; (---), amino acid sequence identical with that of tryptic peptide; (=), polyadenylation sinal.

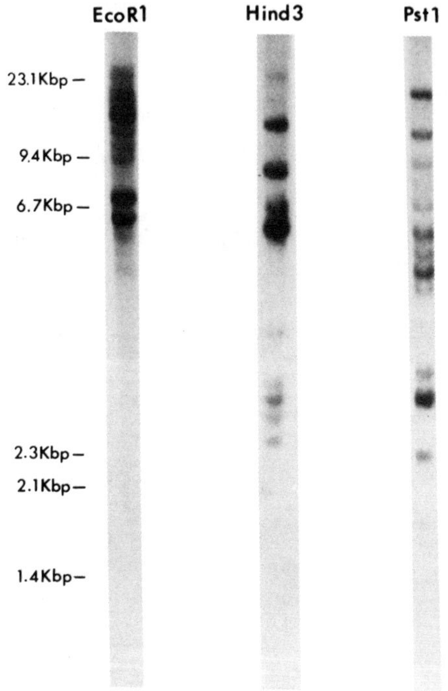

Figure 2. Southern blot analysis of genomic DNA from normal human liver. Hybridization patterns of Eco RI-(lane 1), Hind III-(lane 2), and Pst I-(lane 3)digested DNA to ^{32}P-labeled OAT cDNA probe are shown. Size markers are Hind III-digested λ and Hae III-digested ϕX 174 DNA.

---

temperatures indicated the presence of at least four copies of OAT or OAT-related gene sequences, one of which is the functional OAT gene corresponding to the cDNA (Zintz and Inana, 1989). We have mapped the OAT gene sequences to chromosomes 10 and X, and confirmed that the functional OAT gene is on the former (Fig. 3; Barrett et al., 1987). The chromosomal mapping result appeared to confirm the presence of the OAT gene family and the autosomal recessive inheritance of GA. Cloning and characterization of the functional OAT gene has demonstrated it to be approximately

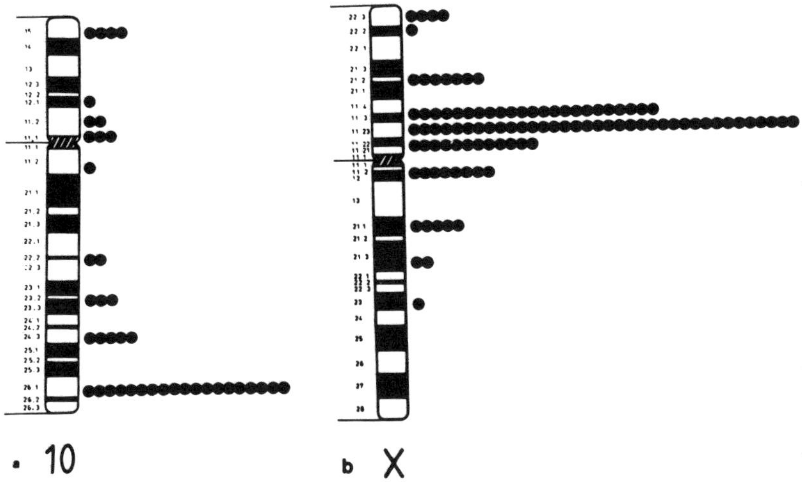

Figure 3. Distribution of silver grains on chomosomes 10 and X. Results of autoradiography of in situ hybridization of radioactively labeled OAT cDNA probe to specific chromosomes are shown.

22 kbp long with 11 exons and a promoter region that has features of both tissue-specific and housekeeping genes (Mitchell et al., 1988a; Zintz and Inana, 1989). The analysis of OAT gene clones also demonstrated at least one processed pseudogene on the X chromosome (data not shown).

We used the OAT probe to investigate the status of the OAT gene and its expression in tissues from GA patients (Inana et al., 1988). High-molecular-weight genomic DNAs of leukocytes or skin fibroblasts from 14 GA patients and normal subjects were subjected to a Southern analysis using the OAT cDNA probe. The results for four representative cases and a control are shown in Figure 4. The hybridization analysis of the OAT gene sequences in the patients identified a case containing a novel 5 Kbp Eco RI genomic fragment, not found in normal subjects or other GA patients (Fig.4, Patient 35). The 5 Kbp fragment appears to be a truncated form of one of the two 5.7 Kbp allelic fragments containing the functional OAT gene sequences

(Barrett et al., 1987) since the latter shows a 50% decrease in relative hybridization intensity. That this case represents a partial heterozygous deletion of the OAT gene was supported by the additional demonstration of a truncated form of a Hind III, Bgl II, Sph I and Msp I functional OAT gene fragment in this patient (data not shown) and the failure to detect any of the altered forms in up to 30 subjects, making it unlikely for them to be restriction fragment length polymorphisms (RFLP).

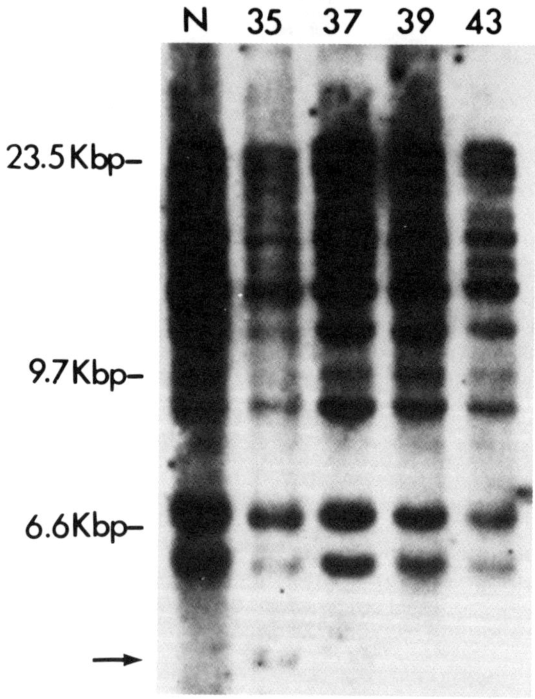

Figure 4. Southern blot analysis of genomic DNA from normal and GA patients. The hybridization patterns of Eco RI OAT genomic fragments are shown for four representative GA patients (35, 37, 39 and 43) and normal control (N). Arrow points to the novel 5 Kbp fragment in patient 35.

RFLPs were observed with Eco RI (6.5 Kbp and 5.7 Kbp fragments), Hind III (14.3 Kbp and 12 Kbp fragments) and Pst I (5.5 Kbp and 5.7 Kbp fragments) in the OAT gene sequences (Fig.5). The approximate frequency of occurrence of the less common variant in each of the RFLP sets was 13% for the 5.7 Kbp Eco RI fragment, 5% for the 12 Kbp Hind III fragment, and 13% for the 5.7 Kbp Pst I fragment among the subjects. The Eco RI polymorphism is present in the functional OAT gene sequence located on chromosome 10 whereas the Hind III and Pst I polymorphisms are present in the OAT-related gene sequences located on the X chromosome, as judged by the previous chromosomal mapping of these gene sequences (Barrett et al., 1987).

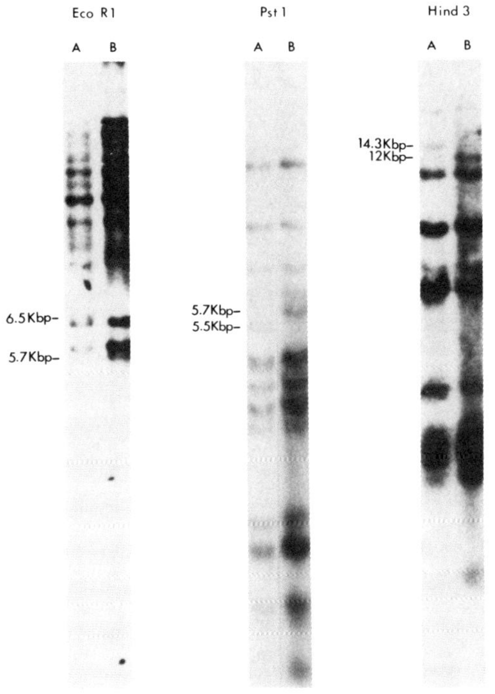

Figure 5. Restriction fragment length polymorphisms of OAT gene sequences. For each enzyme digestion, (A) represents the restriction fragment pattern usually observed, and (B) represents a less common variant containing a restriction fragment length polymorphism (RFLP). RFLPs with Eco RI (6.5 Kbp and 5.7 Kbp fragments), Pst I (5.5 Kbp and 5.7 Kbp fragments), and Hind III (14.3 Kbp and 12 Kbp fragments) are shown.

Skin fibroblasts were obtained from the GA patients, and mRNAs were isolated and subjected to a northern analysis. The results for five representative cases and two controls are shown in Figure 6. Patient 35 is the case with the partial heterozygous deletion of the OAT gene described above. Patient 36 is a B_6-responder, and the rest of the cases are B_6-nonresponders (Weleber et al., 1982). The northern blot analysis of the fibroblast RNAs with the OAT probe indicated the presence of OAT mRNA, similar in size and amount to that found in normal control, in all cases except one (Patient 35), in whom no OAT mRNA was detectable (Fig.6A).

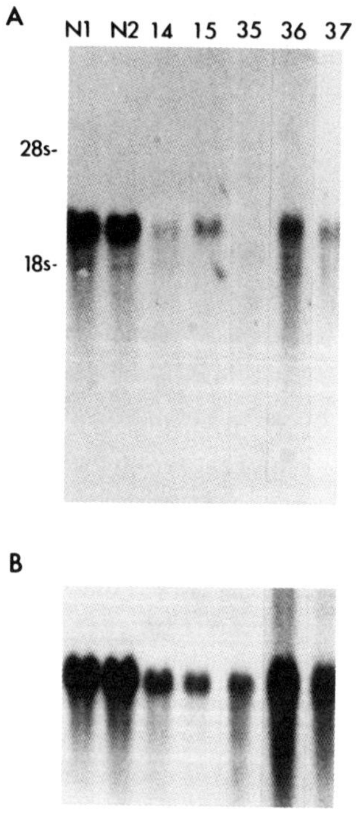

Figure 6. Northern blot analysis of fibroblast RNA from GA patients. (A) OAT mRNA. (B) Actin mRNA. The levels of specific mRNA are shown for five representative GA patients (14, 15, 35, 36 and 37) and normal controls (N1, N2). Position of the 18s and 28s ribosomal RNA are indicated.

The level of OAT protein had been shown to be markedly decreased in patient 14 by immunoassay previously, as mentioned (Ohura et al., 1984). The level of OAT protein present in fibroblasts from patients 35, 36 and 37 was also determined by western blot analysis using the anti-human OAT antibody. The result indicated the presence of variably reduced levels of immunoreactive OAT protein in these cases with patient 35 showing the lowest level, if any, of immunoreactive protein (data not shown).

The finding of an OAT gene, mRNA, and protein defect in one GA case constituted the first real demonstration of the OAT defect in GA at the gene level and provided confirmation for the primary role the OAT gene is thought to play in this disease. The absence of OAT mRNA in patient 35 (patient 4 in Weleber et al., 1982) is most likely due to the partial deletion of one of the OAT gene alleles and a mutation in the regulatory region of the other OAT gene allele which results in nontranscription of the gene and/or a mutation in the gene itself, which in turn results in lability or defective processing and rapid degradation of the transcribed mRNA. Examples of both types of gene mutation have been demonstrated in thalassemia (Orkin and Nathan, 1981).

The results on the status of OAT gene expression in GA patients appeared to correlate well with the clinical heterogeneity that is observed for GA. It is likely that a complete lack of OAT gene expression as essentially found in patient 35 represents those cases that have no detectable OAT activity, nonresponsiveness to B_6 therapy, and a more rapid progression to blindness. Expression of the OAT gene and synthesis of mRNA and a decreased amount of OAT protein as found in patients 14, 15, 36 and 37 may represent those cases that have residual OAT activity (up to 19% of normal), responsiveness to B_6 therapy, and a milder course of the disease.

The case 35 represents a complete loss of expression of both alleles of the OAT gene by different mechanisms resulting in no mRNA. If each allele of the OAT gene contributes half of the total OAT expression, as indicated by the OAT activity in heterozygous carriers, non-expression of one allele should be reflected at the mRNA level in the obligate heterozygous carriers in this patient's family, and this defect should be inherited in a stable manner. To investigate the inheritance of the OAT gene defects in this family, cellular DNA, RNA, and proteins were isolated from skin fibroblasts of the family members of patient 35 and analyzed (Hotta et al., 1989).

Southern blot analysis of the OAT gene in the family members indicated that the patient had inherited the partially deleted allele of the OAT gene, represented by the 5 Kbp Eco RI fragment, from his father and had passed it on to his daughter and a son (Fig.7). Northern blot analysis of the fibroblast RNAs demonstrated the complete absence of OAT mRNA in the patient, as described above, and decreased levels of OAT mRNA in the father, mother, two sons and daughter (Fig.8). Densitometric standardization

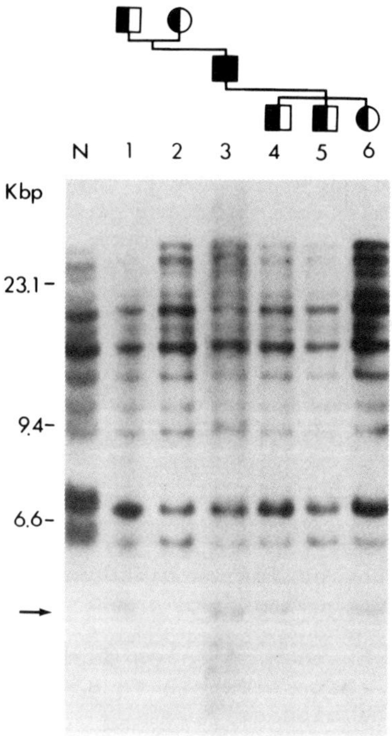

Figure 7. Southern blot analysis of the OAT gene in the GA family members. (N), DNA from normal human liver. (3), fibroblast DNA from the GA patient; (1), his father; (2), mother; (4,5), two sons; and (6), daughter. Eco RI OAT gene fragments are shown. The 6.7, 6.6 and 5.7 Kbp bands represent the functional OAT gene (Barrett et al., 1987). The arrow at 5 Kbp points to the band corresponding to the partially deleted allele of the 5.7 Kbp functional OAT gene band which shows a 50% decrease in hybridization intensity. Size markers are Hind III fragments of bacteriophage λ DNA.

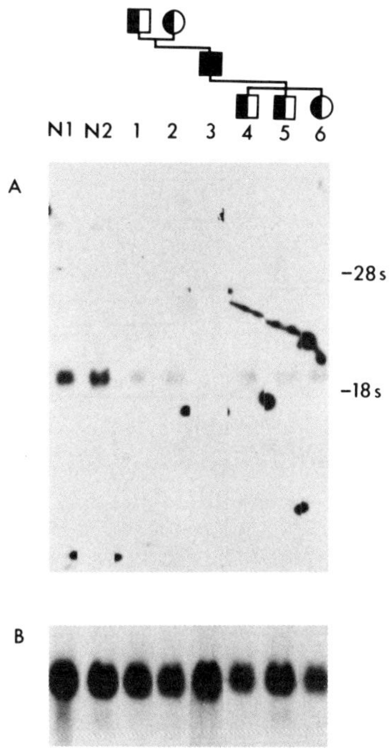

Figure 8. Northern blot analysis of the OAT mRNA in the GA family members. A. Probing for the OAT mRNA with OAT cDNA: (N1,N2), fibroblast RNAs from two normal individuals. (1-6), fibroblast RNAs from the GA family members. Numbers 1-6 are the same as described for Figure 7. B. Probing for the actin mRNA with actin cDNA for standardization of the quality and quantity of RNAs present. Position of the 28s and 18s ribosomal RNA are indicated.

---

of the mRNA levels using the amount of actin mRNA present in each lane indicated that the level of OAT mRNA present in the patient's parents and children, who are all obligate heterozygous carriers, is between 41% and 63% of normal (Table 1). The fibroblast cytosol preparations from the family members were used for assay of OAT activity. The OAT activity in the patient's cells was 3 nmol/hr/mg protein and in the other family members was 170 to 300 nmol/hr/mg protein, which is 35 to 61% of normal (490±98 nmol/hr/mg protein).

Table 1.

Densitometric Analysis of Northern Blot Hybridization

probe	control	1	2	3	4	5	6
		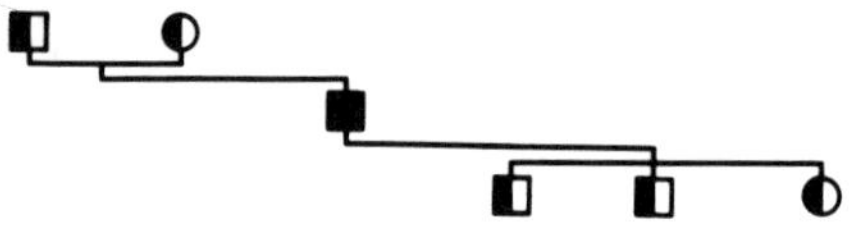					
OAT	1*	0.447	0.353	0	0.393	0.440	0.353
Actin	1	0.994	0.869	0.493	0.670	0.768	0.560
OAT/Actin x100 (%)	100	45.0	40.6	0	58.7	57.3	63.0

* Band intensities on the northern blot were measured by a densitometer, and the areas were calculated. Each value is presented as a ratio to control (normalized to 1) which is an average of two normal human mRNA levels shown in Fig. 2.

Most of the cases with GA examined so far at the molecular level have shown grossly intact OAT genes and mRNAs, suggesting that subtle point mutations are probably the most common defects in GA, as is usually the case in genetic diseases (Ramesh et al., 1986; Inana et al., 1988). The patient 35 is unique in that gross abnormalities, including a partial heterozygous deletion of the OAT gene and a complete absence of the OAT mRNA, are present. Because the molecular genetic defects of OAT are relatively easily identifiable in this patient, the case offered an opportunity to confirm the autosomal recessive inheritance of OAT gene defects in GA in addition to examining the gene dose-product relationship of the OAT gene by examining the status of OAT gene expression in the patient's family members.

The result of the gene analysis identified the patient's father as the source of the partially deleted allele of the OAT gene. The defective allele showed stable inheritance in the family, and the partial deletion could very likely result in inactivation of this allele of the OAT gene as mentioned. The other defective allele coming from the patient's mother was not readily identifiable, but its presence was clearly indicated by the half-normal expression of OAT at both the mRNA and enzyme level. The defect in this allele may involve a point mutation, deletion, or insertion in the gene, especially in the regulatory region, that results in non-expression.

The results of the mRNA and enzyme analysis clearly demonstrated the presence of two distinct heterozygous defects in OAT gene expression in the parents of the patient, their coming together to result in the loss of both functional alleles in the patient, and their transmission to the children to establish the separate heterozygous states. The data demonstrated the stable inheritance of the expression defects of the OAT gene in this family and confirmed the autosomal recessive nature of this disease. The occurrence of two distinct molecular defects, both resulting in lack of expression of the OAT gene in this family, is so far unique among patients with gyrate atrophy.

The data also pointed out that a normal active allele of the OAT gene is responsible for 50% of the total OAT mRNA and enzyme activity in a cell in agreement with the previous demonstration of half-normal level of OAT activity in heterozygous carriers (Trijbels et al., 1977; Valle et al., 1977; O'Donnell et al., 1977). In this regard, the OAT gene allele is similar to many other mammalian gene alleles that show a co-dominant mode of action such as in the hemoglobin gene (Honig and Adams, 1986).

The presence of apparently normal OAT mRNA and a variable amount of immunoreactive OAT protein in most GA patient's fibroblasts, despite the lack of OAT activity in these cells, indicate that a subtle defect, such as a point mutation, is most likely present in the mRNA. The finding of normal-appearing mRNA as seen here is apparently more often the rule in inherited diseases, as demonstrated in patients with ornithine transcarbamylase deficiency (Saheki et al., 1984), argininosuccinate synthetase deficiency (Kobayashi et al., 1986), juvenile form of Sandhoff disease (O'Dowd et al., 1986), partial adenosine deaminase deficiency (Daddona et al., 1985), and thalassemia (Orkin and Nathan, 1981). The finding of a normal pattern of OAT gene sequences in most of the GA patients tested by genomic Southern analysis is also consistent with the likely presence of a subtle mutation, such as a point mutation, instead of gross deletions or rearrangements, in most cases of GA. This has also been found to be the case in patients with ornithine transcarbamylase deficiency (Rozen et al., 1985) and β-thalassemias (Orkin and Nathan, 1981). A subtle mutation in the OAT gene would also be more likely to allow for synthesis of OAT mRNAs from it, as actually demonstrated.

A point mutation in the mRNA may result in poor translation of the message, lability or defective transport of the translated protein to the mitochondria, or inactive OAT protein. In this regard, another case of GA was very informative in elucidating the molecular mechanism of a point mutation defect. This patient was a 31 year old Japanese male with a documented diagnosis of GA based on progressive blindness with the characteristic funduscopic appearance of the retina, hyperornithinemia which is non-responsive to B_6, and undetectable OAT activity in his fibroblasts (Nakajima et al., 1981). There was no history of consanguinity in his family. The concentration of OAT protein in his fibroblasts was shown to be markedly decreased (1.9% of normal) in a previous immunoassay (Ohura et al., 1984). Thus, the basis of this patient's OAT deficiency appeared to be a defect that results in a virtual absence of the OAT protein such as non-expression of the OAT gene or expression of a defective, labile OAT protein.

In order to determine the precise nature of the OAT defect in this patient, high molecular weight DNA was first isolated from his leukocytes and skin fibroblasts and used in a genomic Southern analysis to examine the OAT gene. No gross abnormality could be detected in the Eco RI, Pst I, Hind III, Bgl II, and Sph I patterns of the functional OAT gene, ruling out significant deletions or rearrangements of the gene (data not shown).

Expression of the OAT gene was checked by examining the patient's fibroblast RNA for the presence of OAT mRNA by a northern analysis. The results indicated the presence of OAT mRNA in the fibroblasts, albeit at a decreased level compared to normal (Fig. 9a). A quantitative determination of the OAT mRNA level using the results of actin-probing of the same blot as a normalization standard (Fig. 9b) indicated that the level of OAT mRNA in the patient's fibroblasts is half of that present in normal control fibroblasts (densitometries of the mRNA hybridization intensities of normal and GA with the OAT probe were 15.9 and 4.0, respectively, and with the actin probe were 63.0 and 30.5, respectively; thus, approximation of the ratio of OAT mRNA in GA to normal would be (4.0/15.9) x (63.0/30.5)

= 0.5). In view of the finding that a normal allele of the OAT gene in an obligate heterozygous carrier of GA contributes half of the total OAT mRNA and activity normally found in cells (Valle et al., 1977; Hotta et al., 1989), the results of the northern analysis suggested that in this patient only one of the two alleles of the OAT gene is being expressed.

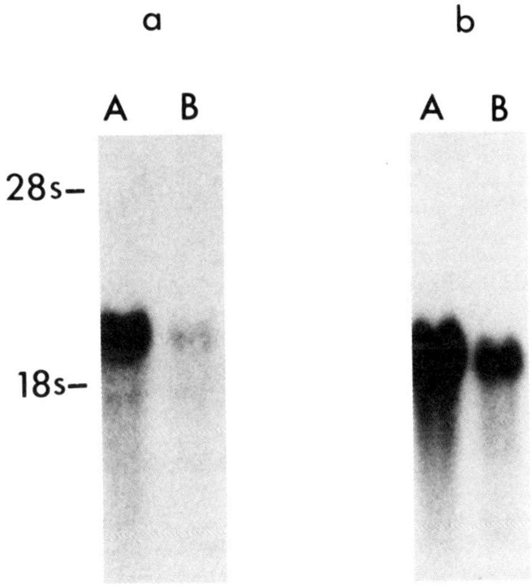

Figure 9. Northern blot analysis of fibroblast RNA from normal (A) and GA patient (B). The levels of specific mRNA (a, OAT mRNA level; b, actin mRNA level) are shown for normal control (A) and the GA patient (B). Ten micrograms (spectrophotometrically determined) of RNA were electrophoresed in each lane, but the actual amount of RNA present in each lane on the blot was determined by normalization with actin mRNA levels (see text). The positions of the 18s and 28s ribosomal RNAs are indicated.

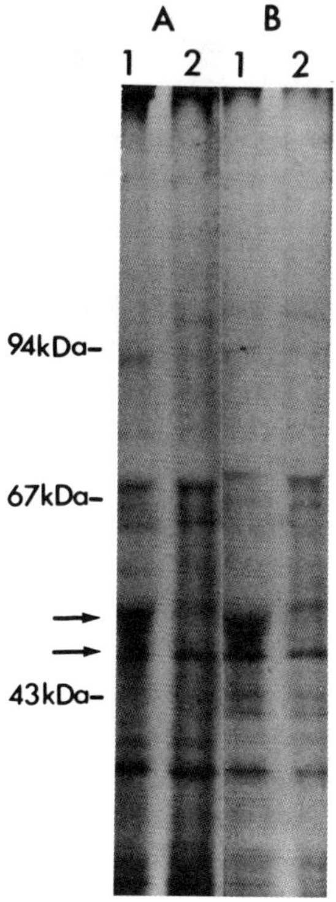

Figure 10. SDS-polyacrylamide gel electrophoresis of ^{35}S-labeled in vitro translation products of normal and GA fibroblast RNA. (A,B) Autoradiography of SDS-polyacrylamide gel electrophoresis of ^{35}S-labeled translation products of normal and GA fibroblast RNAs, respectively, after immunoprecipitation with (1) and without (2) anti-human OAT antibody. The upper arrow points to the 48 kDa immunoprecipitated OAT precursor among the normal and GA translation products. A band just below the precursor may represent a partial processing or degradation of the precursor which may occur in the rabbit reticulocyte lysate in vitro translation system. The lower arrow indicates the position of the purified 45 kDa mature OAT protein in the gel. The molecular size markers are phosphorylase b (94 kDa), bovine serum albumin (67 kDa), and ovalbumin (43 kDa).

The functional status of the expressed OAT mRNA was tested by subjecting the total fibroblast RNA to in vitro translation and immunoprecipitating the protein products with anti-human OAT antibody to detect OAT formation. Analysis of the immunoprecipitated products by SDS-polyacrylamide gel electrophoresis indicated synthesis from the patient's expressed mRNA of an OAT protein similar in size to normal OAT precursor protein (Fig. 10). A slightly smaller immunoprecipitable protein which may be a partially processed or degraded precursor was also observed. Complete processing of the precursor into the mature OAT subunit, however, was not observed in the rabbit reticulocyte lysate system. The results demonstrated that the expressed OAT mRNA is functional in the sense that it is translatable into an immunologically detectable OAT protein.

In view of the finding of grossly intact OAT gene, one expressing allele, and a functional OAT mRNA that can be translated into an OAT protein, it appeared that the marked decrease in the OAT protein found in this patient is most likely due to a subtle change in his expressed OAT that results in a premature disappearance of the protein. In order to investigate this possibility, the patient's fibroblast mRNA was used to prepare a λgt11 cDNA library, and cDNA clones representing the expressed OAT mRNA were isolated using the normal OAT cDNA as a probe. Two overlapping cDNA clones, S1-4a and S1-3b, were isolated which together make up a full-length copy of the patient's OAT mRNA (Fig.11). Sequencing analysis (both strands) of the cDNA clones demonstrated an OAT mRNA sequence identical to normal except for a single point mutation of C to T at residue 1010 resulting in an amino acid codon change of Histidine (CAT) to Tyrosine (TAT) at position 319 (Fig. 12).

The OAT sequence around residue 319 was analyzed for secondary structure and hydrophobicity changes possibly induced by the amino acid change by the Peptidestructure program of the University of Wisconsin Genetics Computer Group (Garnier et al., 1978; Devereux et al., 1984). The analysis indicated an increase in hydrophobicity and flexibility in the region and, significantly, a possible disruption of an α-helical region as a result of the amino acid change (Fig. 13).

Since there is hardly any OAT protein present in this GA patient, it seemed critical to examine the initial phase of the life of this enzyme in this case to see if the newly synthesized precursor can be transported to the mitochondria and processed into a stable mature form. In order to demonstrate that the Histidine to Tyrosine change really does result in a significant functional alteration

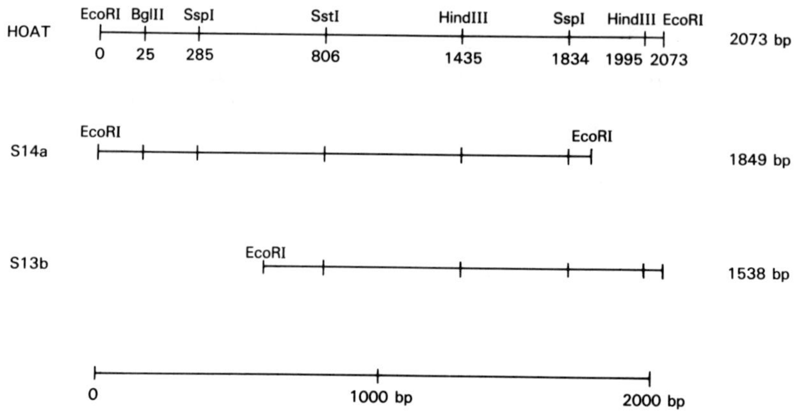

Figure 11. OAT cDNA clones. (HOAT) Normal human OAT cDNA (Inana et al., 1986). (S1-4a, S1-3b) Two cDNA clones representing a full-length copy of the GA patient's expressed OAT mRNA. Major restriction enzyme sites, including the λgt11 cloning site (Eco RI), are shown for each cDNA. A scale bar of 2000 base pairs (bp) is at the bottom.

---

affecting the transport, processing, or stability of OAT, transcriptional expression clones of normal and mutant OAT cDNA were constructed using the pT7/T3-18 vector. In order to examine only the effect of the point mutation, the mutant clone was constructed by replacing a region of the mutation in the normal cDNA with the mutant sequence (Fig. 14). The clones were used to make the normal and mutant OAT mRNAs in an in vitro transcription system (Fig. 15; Melton et al., 1984; Nielsen and Shapiro, 1986), which were then used to make the corresponding ^{35}S-labeled OAT precursors in an in vitro translation system (Fig. 16;

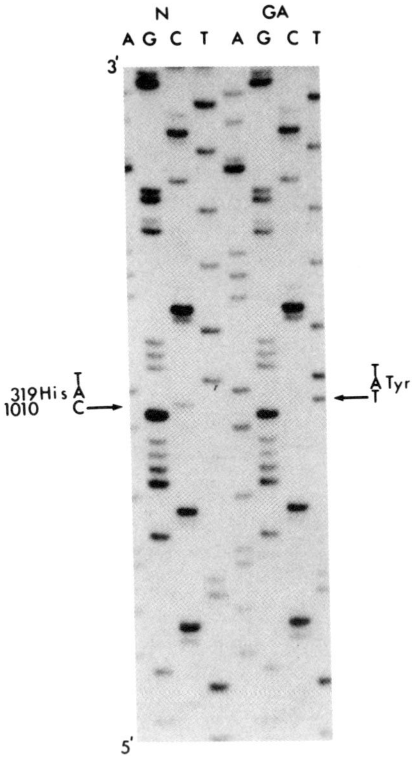

Figure 12. A single-point mutation in the GA patient's OAT mRNA. (N, GA) Normal and GA patient's OAT cDNA sequence in the region of the single-point mutation, respectively. The arrows point to the C (normal) to T (GA) mutation at position 1010 that results in the amino acid codon change from histidine to tyrosine at residue 319. The 5' at the bottom and 3' at the top indicate the 5'to 3' direction of the mRNA. (A,G,C,T) Adenine, guanine, cytosine, and thymine, respectively.

Pelham and Jackson, 1976; Mead et al., 1985). The normal and mutant OAT precursors were tested for mitochondrial transport and processing in an in vitro assay system (Ono and Tuboi, 1986; Conboy and Rosenberg, 1981) and analyzed by immunoprecipitation with anti-human OAT antibody and SDS-polyacrylamide gel electrophoresis. The result indicated that both precursors disappear from the assay media and are transported to the mitochondria (Fig. 17; N and GA, lane 2,), although the transport was less efficient

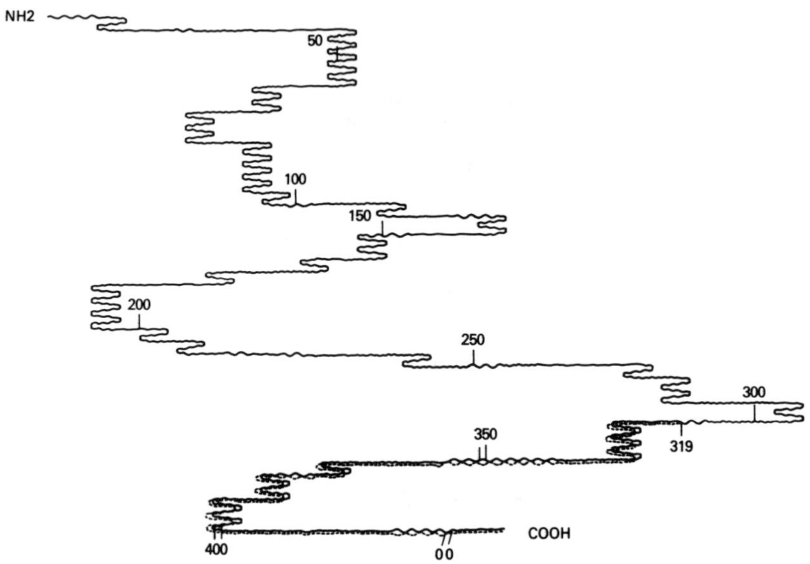

Figure 13. Predicted secondary structure of normal and GA patient's OAT. The secondary structure of normal and GA patient's OAT protein was predicted by the Pepetidestructure program of the University of Wisconsin Genetics Computer Group (Garnier et al., 1978; Devereux et al., 1984). (solid line) normal OAT sequence, (broken line) GA patient's OAT sequence, (zig-zag region) β-sheet, (wavy region) α-helix, (straight line) random coil. Possible premature termination of an α-helical region indicated at position 319 in the GA OAT resulting in a displacement of the distal portion of the protein (broken line).

---

for the mutant precursor. Within the mitochondria the normal precursor was processed to the smaller mature protein (Fig. 17; N, lane 3, lower arrow,), whereas the mutant precursor from the patient was minimally processed, if at all, and remained in the precursor form (Fig. 17; GA, lane 3, upper arrow).

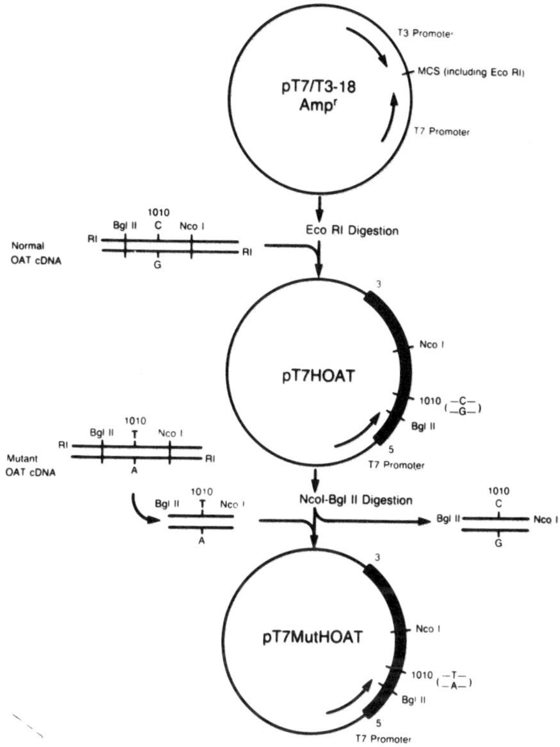

Figure 14. Construction of normal and GA patient's mutant OAT transcriptional expression clones. (pT7/T3-18) the transcriptional expression vector that uses the bacteriophage T7/T3 RNA polymerase promoters for synthesis of RNA. (pT7HOAT) the transcriptional expression clone containing the normal human OAT cDNA with a C-G at position 1010 and read from the T7 promoter. (pT7MutHOAT) the transcriptional expression clone containing the GA patient's mutant OAT cDNA sequence with a T-A at position 1010, also read from the T7 promoter.

Thus, this GA case was an important additional confirmation of OAT gene defect in this disease. The results indicated that most likely only one of the two alleles of the OAT gene is being expressed in this patient and that the expressed OAT contains a single point mutation that results in an amino acid change. In regards to the inactive allele, a complete non-expression of an OAT gene

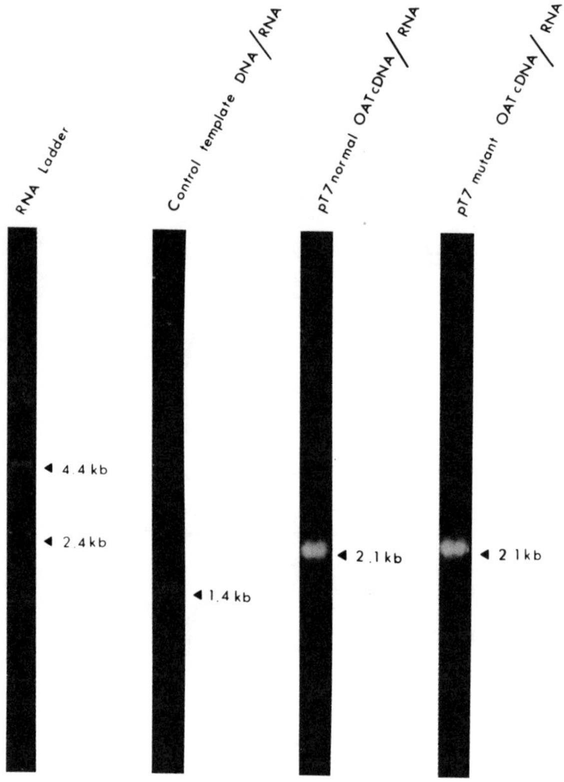

Figure 15. In vitro synthesis of mRNA. pT7/T3-18 vector containing control template DNA, normal human OAT cDNA, and mutant OAT cDNA were used in in vitro transcription system to make the corresponding mRNAs. The mRNAs were analyzed in denaturing agarose gel. RNA size markers were used.

---

allele is similar to what was found in the previous case (Inana et al., 1988) and is probably due to a mutation in the promoter/regulatory region of the gene, which is currently being investigated. In order to explain the marked decrease in the OAT protein level in this patient, it was necessary to examine the nature of his expressed OAT. Within the limitation of DNA sequencing accuracy, the only difference between the normal and GA OAT mRNA sequence, which was obtained from two overlapping clones, was the C to T change at position 1010. The change from Histidine, a charged basic residue, to Tyrosine, an uncharged residue, is a significant mutation. This

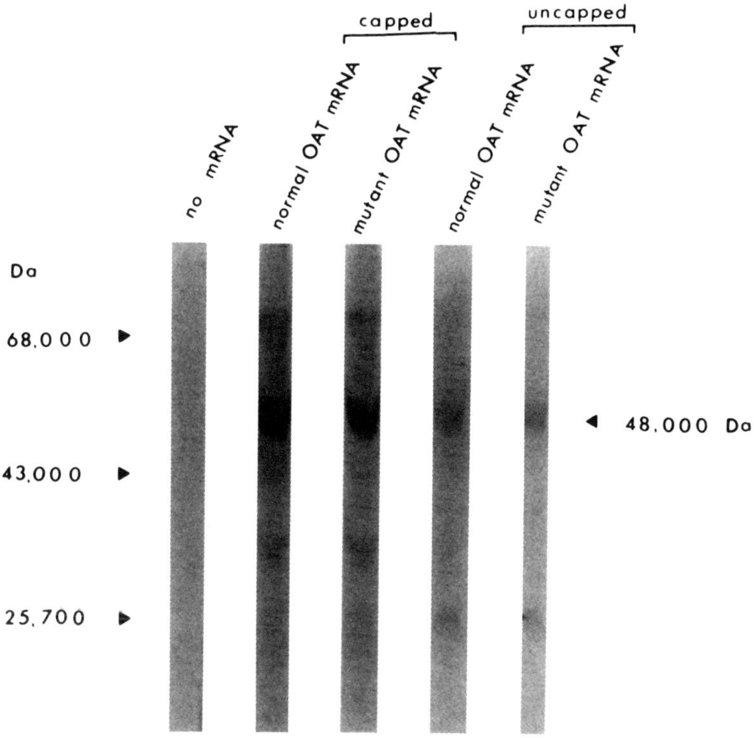

Figure 16. In vitro synthesis of OAT precursors. Normal human and mutant OAT mRNAs, capped and uncapped, were used in in vitro translation system to make the corresponding OAT precursor proteins. The proteins were analyzed in SDS-polyacrylamide gel by electrophoresis. Molecular size markers are bovine serum albumin (68 kDa), ovalbumin (43 kDa), and α-chymotrypsinogen (25.7 kDa).

particular amino acid change has been described for some of the methemoglobins in which the mutation has such a significant effect on the heme-globin bond as to prevent reversible oxygenation of the molecule (Winslow and Anderson, 1983). Even more important than the nature of the specific amino acid change was the possibility that this amino acid change at position 319 may cause a disruption of an α-helical region in the OAT protein. Such a change may cause a significant shift in the three-dimensional orientation of the OAT molecule distal to

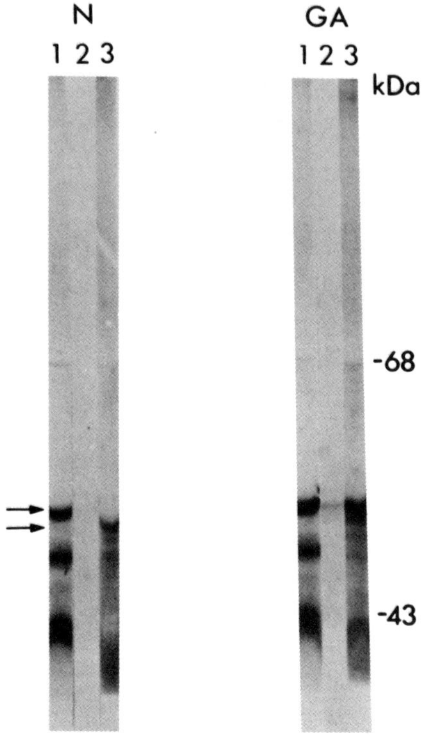

Figure 17. Autoradiography of SDS-polyacrylamide gel electrophoresis of immunoprecipitated, ^{35}S-labeled OAT precursors before and after in vitro mitochondrial transport/processing assay. (N) OAT precursor made from the normal OAT expression clone, pT7HOAT. (GA) OAT precursor made from the GA patient's mutant OAT expression clone, pT7MutHOAT. (1) ^{35}S-labeled OAT precursor before the mitochondrial assay. (2) the supernatant fraction after incubation with mitochondria. (3) the mitochondrial pellet fraction after incubation with mitochondria. (upper arrow) position of the 48 kDa OAT precursor protein. (lower arrow) position of the 45 kDa processed, mature OAT protein. Molecular size markers are bovine serum albumin (68 kDa) and ovalbumin (43 kDa).

---

position 319 since the continuation of the distal portion of the protein may start at a different point in space. That the Histidine residue at position 319 may play such an important role in the structure of the protein was suggested by the conservation of this residue in the human, rat and yeast OAT (Mitchell et al., 1988a). The definitive proof that this amino acid change does result in a significant functional alteration of the OAT protein came through expression of the normal and mutant OAT precursors and assay for mitochondrial transport/processing in an in vitro system. Most of the patient's mutant OAT precursor could be transported to the mitochondria but was only minimally processed there. This would presumably lead to the labile OAT precursor eventually becoming degraded, resulting in a very low level of the OAT protein and activity as seen in this patient.

Demonstration of a single-point mutation in OAT which accounts for the OAT defect present in this patient not only confirmed the molecular genetic basis of the OAT defect in gyrate atrophy, but also raised the level of molecular elucidation of this genetic ocular disease to that achieved for other notable genetic diseases such as hemoglobinopathies (Winslow and Anderson, 1983), α-thalassemia (Goossens et al., 1982), and Gaucher's disease (Tsuji et al., 1987). This GA case with a markedly decreased level of OAT containing an amino acid change contrasted with our previously documented case of GA with a complete absence of the OAT mRNA and a partial heterozygous deletion of the OAT gene (Inana et al., 1988). The clinical severity of the disease in this patient, including the B_6-non-responsiveness, appeared to correlate with the nature of the demonstrated molecular genetic defects of OAT, which are non-expression of one allele and a point mutation resulting in non-processing and virtual absence of the enzyme in the other allele. Along with the recently published results of Mitchell et al. (1988b) and Ramesh et al. (1988), our results indicated that many cases of GA are compound heterozygotes, reflecting the molecular genetic heterogeneity of this disease. The observed heterogeneity at the molecular genetic level may in fact be the basis of the well known clinical heterogeneity seen in this disease (Kaiser-Kupfer et al., 1980).

REFERENCES

Barrett DJ, Bateman JB, Sparkes RS, Mohandas T, Klisak I, Inana G (1987). Chromosomal localization of human ornithine aminotransferase gene sequences. Invest Ophthalmol Vis Sci 28:1037-1042.

Conboy JG, Rosenberg LE (1981). Posttranslational uptake and processing of in vitro synthesized ornithine aminotranscarbamoylase precursor by isolated rat liver mitochondria. Proc Natl Acad Sci USA 78:3073-3077.

Daddona PE, Davidson BL, Perignon JL, Kelley WN (1985). Genetic expression in partial adenosine deaminase deficiency. J Biol Chem 260:3875.
Deshmukh DR, Srivastava SK (1984). Purification and properties of ornithine aminotransferase from rat brain. Experientia 40:357-359.
Devereux J, Haeberli P, Smithies O (1984). A comprehensive set of sequence analysis programs for the VAX. Nucleic Acids Res. 12:387-395.
Garnier J, Osguthorpe DJ, Robson B (1978). Analysis of the accuracy and implications of simple methods for predicting the secondary structure of globular proteins. J Mol Biol 120:97-120.
Goossens M, Lee KY, Liebhaber SA, Kan YW (1982). Globin structural mutant $\alpha^{125Leu \rightarrow Pro}$ is a novel cause of α-thalassaemia. Nature 296:864-865.
Graf-Hausner U, Wilson KJ, Christen P (1983). The covalent structure of mitochondrial aspartate aminotransferase from chicken. J Biol Chem 258:8813-8826.
Honig RH, Adams JG (1986). "Human Hemoglobin Genetics." New York: Springer-Verlag/Wien.
Hotta Y, Kennaway NG, Weleber RG, Inana G (1989). Inheritance of ornithine aminotransferase gene, mRNA and enzyme defect in a family with gyrate atrophy of the choroid and retina. Am J Hum Genet 44:353-357.
Inana G, Totsuka S, Redmond M, Dougherty T, Nagle J, Shiono T, Ohura T, Kominami E, Katunuma N (1986). Molecular cloning of human ornithine aminotransferase mRNA. Proc Natl Acad Sci (USA) 83:1203-1207.
Inana G, Hotta Y, Zintz C, Takki K, Weleber RG, Kennaway NG, Nakayasu K, Nakajima A, Shiono T (1988). Expression defect of ornithine aminotransferase gene in gyrate atrophy. Invest Ophthalmol Vis Sci 29:1001-1005.
Kaiser-Kupfer MI, Valle D, Bron AJ (1980). Clinical and biochemical heterogeneity in gyrate atrophy. Am J Ophthalmol 89:219-222.
Kobayashi K, Saheki T, Imamura Y, Noda T, Inoue I, Matuo S, Hagihara S, Nomiyama H, Jinno Y, Shimada K (1986). Messenger RNA coding for argininosuccinate synthetase in citrullinemia. Am J Hum Genet 38:667.
Mead DA, Skorupa ES, Kemper B (1985). Single-stranded DNA SP6 promoter plasmids for engineering mutant RNAs and proteins: Synthesis of a 'stretched' preparathyroid hormone. Nucleic Acids Res 13:1103-1118.
Melton DA, Krieg PA, Rebagliati MR, Maniatis T, Zinn K, Green MR (1984). Efficient in vitro synthesis of biologically active RNA and RNA hybridization probes from plasmids containing a bacteriophage SP6 promoter. Nucleic Acids Res 12:7035-7056.

Mitchell GA, Looney JE, Brody LC, Steel G, Suchanek M, Engelhardt JF, Willard HF, Valle D (1988a). Human ornithine-δ-aminotransferase: cDNA cloning and analysis of the structural gene. J Biol Chem 263:14288-14295.

Mitchell GA, Brody LC, Looney J, Steel G, Suchanek M, Dowling C, Kaloustian VD, Kaiser-Kupfer M, Valle D (1988b). An initiator codon mutation in ornithine-δ-aminotransferase causing gyrate atrophy of the choroid and retina. J Clin Invest 81:630-633.

Mueckler MM, Himeno M, Pitot HC (1982). In vitro synthesis and processing of a precursor to ornithine aminotransfease. J Biol Chem 257:7178-80.

Mueckler MM, Pitot HC (1985). Sequence of the precursor to rat ornithine aminotransferase deduced from a cDNA clone. J Biol Chem 260:12993-12997.

Nakajima H, Hayasaka S, Shiono T, Watanabe S, Mizuno K, Saito T, Tada K, Watanabe S, Ohba N (1981). A case of gyrate atrophy of the choroid and retina associated with hyperornithinemia. Jpn J Ophthalmol 25:495-500.

Nielsen DA, Shapiro DJ (1986). Preparation of capped RNA transcripts using T7 RNA polymerase. Nucleic Acids Res 14:5936.

O'Donnell JJ, Sandman RP, Martin SR (1977). Deficient L-ornithine: 2-oxoacid aminotransferase activity in cultured fibroblasts from a patient with gyrate atrophy of the retina. Biochem Biophys Res Commun 79:396-399.

O'Dowd BF, Klavins MH, Willard HF, Gravel R, Lowden JA, Mahuran DJ (1986). Molecular heterogeneity in the infantile and juvenile forms of Sandhoff Disease (o-variant G μ2 gangliosidosis). J Biol Chem 261:12680.

Ohura T, Kominami E, Tada K, Katunuma N (1982). Crystallization and properties of human liver ornithine aminotransferase. J Biochem 92:1785-1792.

Ohura T, Kominami E, Tada K, Katunuma N (1984). Gyrate atrophy of the choroid and retina: decreased ornithine aminotransferase concentration in cultured skin fibroblasts from patients. Clin Chim Acta 136:29-37.

Ono H, Tuboi S (1986). Translocation of proteins into rat liver mitochondria. Eur J Biochem 155:543-549.

Orkin SH, Nathan DG (1981). The molecular genetics of thalassemia. In Harris H, Hirschhorn K (eds): "Advances in Human Genetics," New York: Plenum, pp 233-280.

Pelham HR, Jackson RJ (1976). An efficient mRNA-dependent translation system from reticulocyte lysates. Eur J Biochem 67:247-256.

Peraino C, Pitot HC (1963). Ornithine-δ-transaminase in the rat: I. Assay and some general properties. Biochim Biophys Acta 73:222-231.

Ramesh V, Shaffer MM, Allaire JM, Shih VE, Gusella JF (1986). Investigation of gyrate atrophy using a cDNA clone for human ornithine aminotransferase. DNA 5:493-501.
Ramesh V, McClatchey AI, Ramesh N, Benoit LA, Berson EL, Shih VE, Gusella JF (1988). Molecular basis of ornithine aminotransferase deficiency in B-6-responsive and -non-responsive forms of gyrate atrophy. Proc Natl Acad Sci USA 85:3777-3780.
Rozen R, Fox J, Fenton WA, Horwich AL, Rosenberg LE (1985). Gene deletion and restriction fragment length polymorphisms at the human ornithine transcarbamylase locus. Nature 313:815.
Saheki T, Imamura Y, Inoue I, Miura S, Mori M, Ohtake A, Tatibana M, Katsumata N, Ohno T (1984). Molecular basis of ornithine transcarbamylase deficiency lacking enzyme protein. J Inher Metab Dis 7:2.
Sanada Y, Sunemori I, Katunuma N (1970). Properties of ornithine aminotransferase from rat liver, kidney and small intestine. Biochim Biophys Acta 220:42-50.
Shiotani T, Sanada Y, Katunuma N (1977). Studies on the structure of rat liver ornithine aminotransferase. J Biochem 81:1833-1838.
Simell O, Takki K (1973). Raised plasma ornithine and gyrate atrophy of the choroid and retina. Lancet 1:1030-1033.
Trijbels JMF, Sengers RCA, Bakkeren JAJM, DeKort AFM, Deutman AF (1977). L-ornithine-ketoacid-transaminase deficiency in cultured fibroblasts of a patient with hyperornithinemia and gyrate atrophy of the choroid and retina. Clinica Chimica Acta 79:371-377.
Tsuji S, Choudary PV, Martin BM, Stubblefield BK, Mayor JA, Barranger JA, Ginns EI (1987). A mutation in the human glucocerebrosidase gene in neuronopathic Gaucher's disease. N Engl J Med 316:570-575.
Valle D, Kaiser-Kupfer MI, DelValle LA (1977). Gyrate atrophy of the choroid and retina: Deficiency of ornithine aminotransferase in transformed lymphocytes. Proc Natl Acad Sci (USA) 74:5159-5161.
Valle D, Simell O (1983). The hyperornithinemias: gyrate atrophy of the choroid and retina. In Stanbury JB, Wyngaarden JB, Fredrickson DS, Goldstein JL, Brown MS (eds): "The Metabolic Basis of Inherited Disease", New York: McGraw Hill, pp. 389-396.
Weleber RG, Wirtz MK, Kennaway NG (1982). Gyrate atrophy of the choroid and retina: Clinical and biochemical heterogeneity and response to vitamin B_6. Birth Defects Original Article Series 18:219-30.

Winslow RM, Anderson WF (1983). Hemoglobinopathies. In Stanbury JB, Wyngaarden JB, Frederickson DS, Goldstein JL, Brown MS (eds): "Metabolic basis of inherited disease," New York: McGraw-Hill, pp 1666-1710.

Young RA, Davis RW (1983). Efficient isolation of genes by using antibody probes. Proc Natl Acad Sci (USA) 40:1194-1198.

Zintz C, Inana G (1989). Analysis of the human ornithine aminotransferase gene family. (Submitted for publication).

Molecular Biology of the Retina: Basic and Clinically Relevant Studies, pages 221–240

DIVERSE MUTATIONS LEAD TO INACTIVATION OF THE RETINOBLASTOMA GENE

Eva Y.-H.P. Lee, Shi Huang, Jin-Yuh Shew, Larry A. Donoso* and Wen-Hwa Lee

Department of Pathology, M-012, School of Medicine, University of California, San Diego, CA 92037, * Wills Eye Hospital, Ninth and Walnut Street, Philadelphia, PA 19107

INTRODUCTION

Retinoblastoma (RB) is a childhood disease which occurs in about 1 in 20,000 live births (Vogel, 1979; Shields, 1983). It is a highly malignant but readily treatable tumor of the retinoblasts. Approximately 40% of all cases display the hereditary form of the disease (Matsunaga, 1978; Vogel, 1979). Karyotypic examination of tumor cells has identified deletions of the long arm of chromosome 13 (Balaban et al, 1982). This deletion has also been described in normal fibroblasts from patients with hereditary retinoblastoma (Francke, 1976; Yunis and Ramsay, 1978). Among all chromosome 13 deletions, band 13q14 was commonly involved. Based on analysis of clinical data, Knudson proposed a "two-hit" model for retinoblastoma genesis (Knudson, 1971): in the hereditary form of the disease the first hit is inherited as a germinal mutation and the second hit as a somatic mutation of the retinoblasts; whereas in the nonhereditary form of the disease both mutations occur in the retinoblasts. Based on this hypothesis, the retinoblastoma gene would act in a recessive manner in contrast to the apparent autosomal dominant transmission of the disease. The recessive nature of the RB gene was supported by studies showing the specific loss of heterozygous chromosome 13 markers even in retinoblastomas without visible chromosome alterations (Cavenee et al, 1983). Using a random probe H3-8 located at 13q14 (Lalanda et al, 1984), a homozygous deletion was found in two retinoblastoma tumors (Dryja, 1986). This probe was subsequently used for cloning and the cDNA obtained was identified as the RB gene based on its aberrant expression in retinoblastoma cells (Friend et al, 1986; Lee et al, 1987a; Fung et al, 1987). Studies of the gene product has confirmed the recessive nature of the mutation: no normal gene product has been identified in retinoblastoma cells examined so far. A detailed analysis of the gene structure and molecular genetic studies of mutations in the gene in tumor cells has revealed several diverse mechanisms leading to the inactivation of the RB gene, which is the main focus of this review.

RB GENE STRUCTURE

By screening several human cDNA libraries, a cDNA sequence of 4,757 nucleotides as shown in Figure 1 was cloned. This sequence encodes the full length RB protein $pp110^{RB}$ (Lee et al, 1987b; Friend et al, 1987). The first in-frame Met (nucleotide 139) was defined as the initiation codon based on the following: 1) the molecular weight (MW) of the predicted protein translated from the first Met was closer to the actual MW of the RB gene product determined in vivo (see below); 2) in vitro translation of the cDNA clone yields full-size product only when the first 200 nucleotides were present (Shew, unpublished data); 3) antibodies prepared against a synthetic peptide corresponding to the region between the first Met and the second Met immunoprecipitated $pp110^{RB}$ (Shew, unpublished data). With the first methionine, the predicted RB protein has 928 amino acids and a mass of 106 kd. An unusual stretch of amino acids of poly-Ala and poly-Pro from residue 10 to 29 may explain the apparent MW of 110 kd in

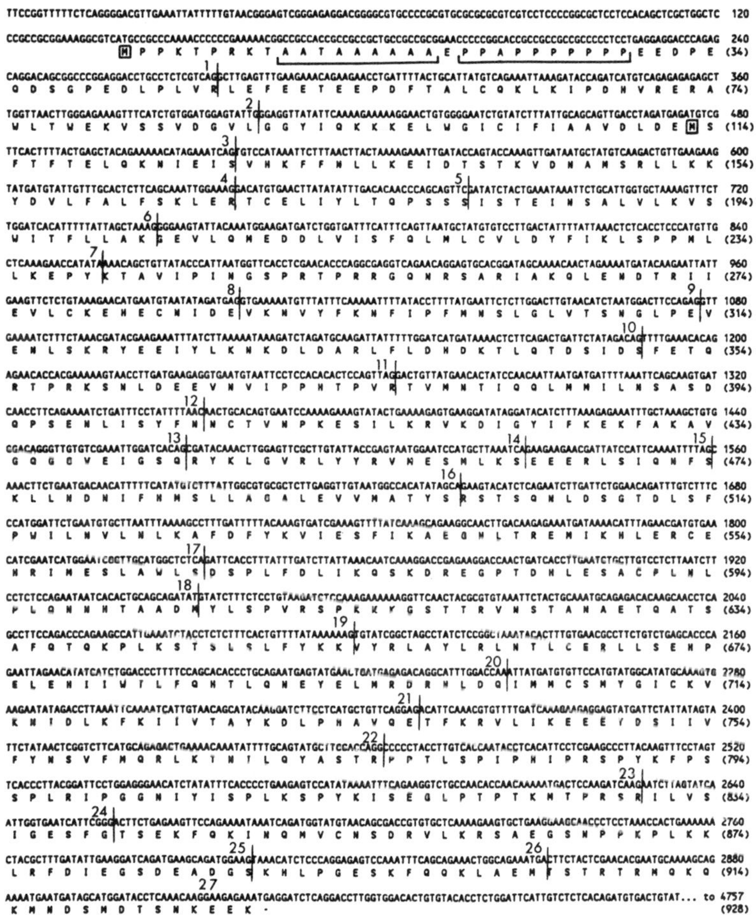

Figure 1. Complete RB cDNA sequence and predicted amino-acid sequence of the RB protein with exon boundaries. (From Lee et al. Nature 329: 642-645, 1987 & Hong et al. Proc. Natl. Acad. Sci USA 86: 5502-5506, 1989).

SDS polyacrylamide gel electrophoresis (SDS-PAGE) since a deletion of this region results in a mutated protein with a much faster mobility than would have been expected (Huang et al, submitted).

Genomic clones of the RB gene spanning 200 kb have been obtained by screening genomic libraries with cDNA probes (Bookstein et al, 1988, Wiggs et al, 1988). A detailed study using a series of oligonucleotides as probes in Southern blot hybridization and as primers for sequencing genomic DNA revealed 27 exons as shown in Figure 1 (Hong et al, 1989; McGee et al, 1989; T'Ang et al 1989). The DNA sequences of all the splice donor and acceptor sites contained the consensus /gt and ag/ intronic dinucleotides. The smallest exon, exon 24, contained only 31 nucleotides. The partial genomic sequence obtained for exon 27, the 3'-terminal exon, revealed that RB cDNA diverged (as AAAAA) from the genomic sequence 16 bp after a consensus AATAAAA polyadenylylation signal (Lee et al, 1987b), indicating that our cDNA clone RB-5 (Lee et al, 1987a) was complete at the 3' end. A major transcription initiation site as determined by S1 nuclease protection (Hong et al, 1989) is located at position -276 with the translation initiation site as +1. No typical TATA box is present near the transcription start site. Localization of the RB promoter region was accomplished by utilizing a heterologous expression system containing a bacterial chloramphenicol acetyltransferase gene (Hong et al, 1989). Deletion analysis revealed that a region as small as 70 bp is sufficient for RB promoter activity, similar to other previously characterized G+C rich promoters. No enhancer element was detected within the 7.3 kb of upstream sequence studied. Thus the RB promoter is very much like that of a "housekeeping" gene, consistent with its ubiquitous expression pattern (Lee et al, 1987a).

THE RB GENE PRODUCT

Using antibodies raised against TrpE-RB fusion proteins, a phosphoprotein of approximately 110 kd ($pp110^{RB}$) was detected in normal cells but not in cultured cells from five of five retinoblastomas (Lee et al, 1987b). Characterization of additional retinoblastoma tumors has not detected the presence of a normal RB protein in any retinoblastoma examined, confirming earlier hypothesis of the recessive nature of the mutated RB gene. Further characterization of $pp110^{RB}$ showed that the RB protein migrated in SDS-PAGE as diffuse bands of MW 110-114 kd (Figure 2, lane 1). After potato acid phosphatase treatment, the complex banding pattern was reduced to a single fast moving band (Figure 2, lane 2). The slower migrating forms were strongly labelled with ^{32}P, suggesting that the apparent size heterogeneity was due to variable phosphorylation (Ludlow, 1989; Shew et al, 1989). $pp110^{RB}$ is primarily located in the cell nucleus as defined by cellular fractionations and immunocytochemical studies (Lee et al, 1987b). Consistent with its nuclear localization and hypothetical regulatory function, $pp110^{RB}$ was found bind to DNA-cellulose (Lee et al, 1987b). Using E.coli expressed TrpE-RB fusion proteins encompassing various regions of the RB gene, we have further localized the DNA binding activity to the C-terminus of $pp110^{RB}$ containing exon 19-27 (Wang et al, submitted).

Besides binding to DNA, the other identified biochemical activity of $pp110^{RB}$ is its complex formation with a number of oncogene products of the DNA tumor viruses, namely E1A protein of adenovirus, large T antigen of SV40, and E7 protein of the papilloma virus (Whyte et al, 1988; DeCaprio et al, 1988; Dyson et al, 1989). The regions of these oncogene products involved in binding to $pp110^{RB}$ coincide with their previously identified transforming regions. It was therefore postulated that association with tumor suppressor gene products such as $pp110^{RB}$ and other cellular proteins may provide a mechanism for the transforming function of these oncoproteins (DeCaprio et al, 1988; Whyte et al, 1989). To help understand the functional significance of these complex formations we have determined the regions of pp110^{RB} involved in binding to T (Huang et al, submitted). A number of deletion mutants generated by in vitro transcription and translation were assayed for their binding activities to large T antigen in vitro. Two distinct regions were found necessary for the binding, region 1 from residues 393 to 572 and region 2 from 648 to 774. Work from the laboratory of E. Harlow (Cold Spring Harbor Laboratory) also showed that the same regions are

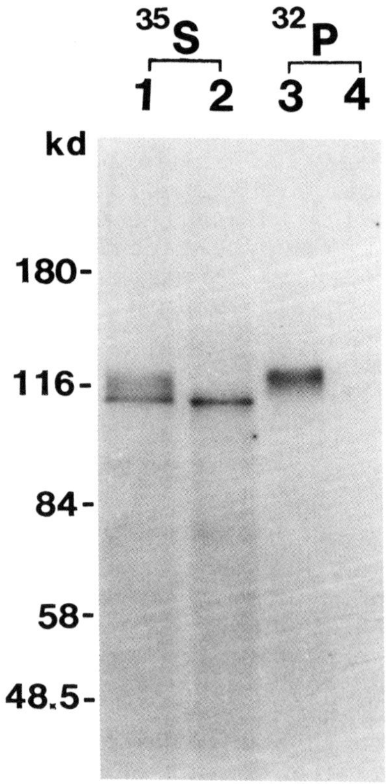

Figure 2. Dephosphorylation analysis of pp110^{RB}. Cells were labeled with either ^{35}S-methionine or ^{32}P-orthophosphate. RB protein in the cell lysate was immunoprecipitated with rabbit anti-RB IgG. The immunoprecipitates were treated with potato acid phosphatase (lanes 2 and 4) or reaction buffer only (lanes 1 and 3) prior to separation by 7.5% SDS-PAGE. (From Shew et al Oncogene Res, 4:205-214).

involved for both E1A and T binding (personal communications). It is of interest to note that relatively large regions of $pp110^{RB}$ are involved in binding to a relative small region of T (amino acids 105 to 114) and EIA (amino acids 30 to 60 and amino acid 121 to 127) (DeCaprio et al, 1988; Whyte et al, 1989). It is likely that a specific conformation of RB protein is required for interaction with the oncoprotein since small insertional mutations abolish the binding activity (Huang et al, submitted). As will be discussed below a few naturally occurring mutants of RB protein characterized so far all involve alterations of these two regions.

MUTATIONAL ALTERATIONS OF THE RB GENE IN HUMAN TUMORS OR CELL LINES

The RB gene was initially identified on the basis of its altered expression in retinoblastoma tumors, as compared to normal retina, placenta, and nonretinoblastoma tumors. Using Southern and Western blotting analysis, we and others have found gross rearrangement of the RB gene in 15%-30% of the retinoblastoma cells examined (Friend et al, 1986; Lee et al, 1987a; Fung et al, 1987). Results from Gallie's group using RNase protection assay suggested that point mutations can be detected frequently in retinoblastoma cell lines with normal-sized mRNAs. Target for the point mutation varied among tumors (Dunn et al, 1988). These studies demonstrated that many kinds of mutations including deletions and point mutations could lead to inactivation of the RB gene. Detailed studies of mutational mechanisms (see below) further illustrate this viewpoint.

It was observed in clinical studies that patients with hereditary retinoblastoma had a higher risk of developing other second primary nonocular tumors, indicating a possible role of RB gene in the development of other tumors types (Abramson et al, 1979; Draper et al, 1986). Therefore many additional tumors have been examined for possible RB gene alterations. Some of the mutations characterized are shown in Table 1. For example, RB mutations has been identified in

TABLE I. Summary of cell lines and tumors studied for RB gene mutations

Tumor Types	RB DNA	RB mRNA	RB protein
Retinoblastoma			
Y79	het del	s	-
RB355	het del	s	-
WERI-1	com del	-	-
WERI-24	hem del	-	-
WERI-27	het del	s	-
Osteosarcoma			
Saos-2	het del	s	95 kd
OHS	het del	s	-
Synovial sarcoma			
SDSY-1	int dup	l	ND
Breast tumor			
MDA-MB 436	int dup	l	-
MDA-MB 468	par del	-	-
Small cell lung carcinoma			
H209	+	+	110 kd
SCLC-SD-1	del	s	104 kd
12 lines	+ or del	+ or -	-
Prostate carcinoma			
DU145	+	s	104 kd
Leukemia			
HSB-2	par del	s	75 kd

Abbreviations: +, gross normal and mutations yet to be defined; -, absent; ND, not determined; het del, heterozyous deletion; com del, complete deletion; int dup, internal duplication; par del, partial deletion; s, shortened transcript; l, lengthened transcript.

approximately 50% of osteosarcoma (Toguchida et al, 1988; Shew et al, 1989). One of the well characterized osteosarcoma cell line, Saos-2, synthesizes a truncated RB protein of MW 95 kd (Shew et al, a, in press). A lengthened mRNA resulted from rearrangement of the RB gene was identified in one primary synovial sarcoma (Mendoza et al, 1988; Shew et al, unpublished data). Partial DNA deletions and absent or altered mRNA were found in two breast tumor cell lines (Lee et al, 1988a). Mutations in additional tumor lines and primary breast tumors have been reported subsequently (T'Ang et al, 1988; Varley et al, 1989). Altered RB gene expression have also been found in small cell lung carcinomas (Harbour et al, 1988; Yokota et al, 1988; Shew et al, b, in press; Hensel et al, submitted). In one primary culture of a small cell lung carcinoma, a mutated RB proteins has been found (see below); in one established cell line, an apparent normal size but under-phosphorylated RB protein was identified (Hensel et al, submitted). Mutated protein products were also observed in one prostate carcinoma DU145 (Bookstein et al, submitted), one bladder carcinoma J-82 (Horowitz et al, 1989) and one leukemia cell line HSB-2 (Cheng et al, submitted). These preliminary results suggest a role for mutations of the RB gene in the genesis of cancers other than retinoblastoma. Taking together with the ubiquitous expression of the RB gene and its tumor suppression activities in different cell types (Huang et al, 1988; Bookstein et al, submitted; Lee et al, unpublished), we speculate that RB gene may have common cellular functions essential for normal growth of many, if not all, cells (for review, see Lee et al, 1989a ;1989b).

MECHANISMS OF RB GENE MUTATIONS

To better understand mechanisms for inactivation of the RB gene, mutations affecting the RB gene in several tumor cell lines have been examined in detail. These studies includes retinoblastoma Y79 (Bookstein et al, 1988; Lee et at, 1988b), bladder carcinoma J82 (Horowitz et al, 1988), prostate carcinoma DU145 (Bookstein et al, submitted), osteosarcoma Saos2 (Shew et al, a, in press), small cell lung carcinoma SCLC-SD (Shew et al, b, in press) and leukemia cell line HSB-2 (Cheng et al, submitted) (Figure 3). Described below is a brief summary of the different mechanisms involved in each cell line.

Figure 3. Structure of mutated RB polypeptides detected in several human tumor cell lines. A schematic representation of RB protein consisting 27 exons is shown above representations of mutated RB protein. RB proteins are shown as open bars and gaps in between denote the deleted regions.

The retinoblastoma cell line Y79 expresses a shortened RB mRNA transcript, however, the RB DNA appears grossly normal (Lee et al, 1987a; Fung et al, 1987). Cloning and sequencing of the Y79 RB cDNA led to the identification of a deletion of 470 bp nucleotides corresponding to exons 2-6 of the wild type DNA (Lee et al, 1988b). The deletion resulted in reading frame change and the premature termination of the translated protein. A possible mechanism involving the RB gene deletion in Y79 was further explored by sequencing the cloned deletion junction as well as corresponding regions in the normal gene (Lee et al, 1988b). Sequences of the deletion endpoints demonstrated no apparent homology to each other or to the Alu family; therefore, the deletion cannot be explained by recombination between homologous sequences. Characterization of the RB gene in Y79 demonstrated one RB allele had a deletion of about 50 kb that removed exons 2-6. The DNA rearrangement was heterozygous because other Y79 genomic clones contained these exons. Since normal-sized mRNA and $pp110^{RB}$ were absent from Y79 cells, the gross normal allele must also be inactivated by mechanisms different from that of the allele containing deletion (Bookstein et al, 1988). A mutation of the promoter is considered most plausible, and our preliminary results have identified a point mutation in the RB promoter region of certain tumor cells (Rio et al, unpublished data).

The bladder carcinoma J82 cell line was initially found lacking $pp110^{RB}$ in an assay for E1A association (Horowitz et al, 1989). However immunoprecipitation with several antibodies against $pp110^{RB}$ all precipitated an RB-related protein of smaller size. To assay the nature of the genetic defect that underlay the alterations of the RB protein in J82 cells, RNA of J82 cell was examined by means of reverse transcription and polymerase chain reaction (Rappolee et al, 1988). A deletion of 105 nucleotides from exon 21 was identified. This "in-frame" deletion results in the elimination of 35 amino acid residues, from positions 697 to 731, but allowed continued translation of distal sequences. All other regions of RNA appeared normal in size in parallel PCR reactions with various synthetic primers. Examination of J82 genomic DNA revealed no gross alterations indicating that the absence of the exon in the mRNA was not due to deletion of corresponding chromosomal sequences. Therefore a detailed study around exon 21 and flanking intron sequences was performed by direct sequencing the PCR-amplified fragments. This revealed a point mutation in the splice acceptor in intron 20. No PCR fragments encoding wild-type splice acceptor sequences were recovered, indicating that J82 cells are either homozygous or hemizygous for the mutant allele (Horowitz et al, 1989).

Osteosarcoma cell line Saos-2 expresses a truncated RB protein of 95 kD that was located mainly in the cytoplasm (Shew et al, a, in press). This protein was identified by antibodies recognizing several different RB epitopes, but not by one directed against the C-terminus, suggesting C-terminal truncation. Analysis of mRNA and genomic DNA revealed that a transcriptionally active RB allele had a deletion of exons 21-27. To understand the deletion mechanism the DNA fragments containing the potential recombination junction from both normal and Saos-2 genomes were cloned and sequenced. No repetitive units such as Alu or KpnI sequence were found in the recombination junction region. The neoplastic phenotype of Saos-2 cells has been shown to be repressed by expressing an exogenous normal RB gene, suggesting that the endogenous p95RB is functionally inactive (Huang et al, 1988). Thus the C-terminal region is likely to be important for normal RB activity. Consistent with this notion, this mutated protein lost all the known RB properties such as complex formation with SV40 large T antigen, posttranlational modification by phosphorylation, and nuclear localization (Shew et al, a, in press).

SCLC-SD-1 is a primary culture of small-cell lung carcinoma. This cell expressed a mutated RB protein of 104 kD (p104). To decipher the mutation involved, RB mRNA was analyzed by the procedure of reverse transcription and PCR amplification (RT-PCR). Direct DNA sequencing of fragments from RT-PCR revealed the absence of the exon 16 sequence. Analysis of genomic DNA from this tumor showed that exon 16 and its flanking splice donor and acceptor sequences were present and entirely normal; however, a 43 bp region containing the splice donor site of intron 15 was deleted instead. Exon 15 was joined directly to exon 17 during mRNA processing via a cryptic splice donor site; exon 16 was presumably skipped because the preceding mutated intron was of insufficient length (<80 bp) for normal RB RNA processing. It is of interest to note that this mutated protein also was unphosphorylated and unable to bind SV40 large T antigen (Shew et al, b, in press).

A large homozygous deletion of exon 18 to beyond exon 27 of the RB gene was found in the leukemia cell line HSB-2. A mutated protein of 72 kd was detected with ^{35}S-methionine labeling but not with ^{32}P-orthophosphate labeling. This deletion would remove >85 kb DNA at the RB locus. A detailed mechanism explains the deletion of this large stretch of DNA remaines to be clarified (Cheng et al, submitted).

PERSPECTIVE

Detailed studies of RB gene defects in tumor cells have uncovered many different mechanisms to inactivate the RB gene. At the DNA level, deletion, insertion and point mutations have been observed in the RB gene. These mutations manifest various effects on RB protein, leading to either a complete absence, or mutated forms, of the protein. Mutant RB proteins are frequently found to be unphosphorylated. The phosphorylation status of $pp110^{RB}$ oscillates during the cell cycle and hyperphosphorylation of $pp110^{RB}$ is correlated with mitogenic stimulation (Chen et al., 1989; DeCaprio et al., 1989; Buchkovich et al., 1989). We speculate that phosphorylation plays an important regulatory role for the function of the RB gene. Other biochemical properties such as complex formation and DNA binding activity are also frequently jeopardized. Using the complementary approach to generate mutant RB proteins in vitro for biochemical analysis, we have defined two functional regions for the formation of complexes with oncoprotein and regions for DNA binding. Since mutations resulting in changes in these regions are frequently found in human tumor cells, we speculate that the regions defined in vitro have important functions in vivo as well.

We have hypothesized that RB protein plays a regulatory role either in cell proliferation or differentiation (for reviews, see Lee et al, 1989a and 1989b). Two areas of studies need further investigations: characterization of the biochemical functions of the RB protein and continued studies of the biological effects of RB expression. Whether RB protein interacts with other cellular proteins for its cellular function and whether it recognizes specific DNA sequence are immediate questions to address. In terms of biological effects, a detailed comparison between RB non-expressing cells and RB expressing cells after the replacement of RB protein via viral infection will be important in addressing the role of RB gene in oncogenesis.

ACKNOWLEDGEMENTS

We thank our colleagues who have contributed to the studies described in this report, Drs. Rob Bookstein, Peter Scully, Huei-Jen S. Huang, Pascale Rio, Frank Hong, Phang-Lang Chen, Nan-Ping Wang and Hoang To. These studies were supported by grants from National Institutes of Health (CA 49649) and from Tobacco Council for Research (#2491) to E.Y-H.P.L.; National Eye Institute (EY07737) to L.A.D.; and National Eye Institute (EY 05758) and the American Cancer Society grant (CD 373) to W.-H.L. S.H. is a recipient of a postdoctoral fellowship from the Cancer Research Coordination Committee, University of California.

REFERENCES

Abramson DH, Ronner HJ, Ellsworth RM. (1979) Second tumors in nonirradiated bilateral retinoblastoma. Am J Opthalmology 87:624-627.

Balaban G, Filbert F, Nichols W, Meadows AT, Shields J. (1982) Abnormalities of chromosome #13 in retinoblastoma from individuals with normal constitional karyotypes. Cancer Genet and Cytogenet 6:213-221.

Bookstein R, Lee EYHP, To H, Young LJ, Sery T, Hayes R, Friedmann T, Lee WH. (1988) Human retinoblastoma susceptibility gene: genomic organization and analysis of heterozyous intragenic deletion mutants. Proc Natl Acad Sci USA 85: 2210-2214.

Bookstein R, Lee WH. Replacement of a mutated retinoblastoma gene suppresses the tumorigenic potential of human prostate carcinoma cells. (submitted)

Buchkovich K, Duffy LA, Harlow, E. (1989) The retinoblastoma protein is phosphorylated during specific phases of the cell cycle. Cell 58:1097-1105.

Cavenee WK, Dryja TP, Phillips RA, Benedict WF, Godbout R, Gallie BL, Murphree AL, Strong LC, White RL. (1983) Expression of recessive alleles by chromosomal mechanisms in retinoblastoma. Nature 305:779-784.

Chen PL, Scully P, Shew JY, Wang, JYJ and Lee, WH. (1989) Phosphorylation of the retinoblastoma gene product is modulated during the cell cycle and cellular differentiation. Cell 58:1193-1198.

Cheng J, Scully P, Shew J, Lee WH, Villa V and Haas M. Homozygous deletion of the retinoblastoma gene in an acute lymphoblastic leukemia (T) cell line. (submitted)

DeCaprio JA, Ludlow JW, Figge J, Shew J-Y, Huang C-M, Lee W-H, Marsillo E, Paucha E, Livingston D. (1988) SV40 large tumor antigen forms a specific complex with the product of the retinoblastoma susceptibility gene. Cell 54:275-283.

DeCaprio, JA, Ludlow, JW, Lynch, D, Fuukawa, Y, Griffin, J, Piwnica-Worms H, Huang CM, Livingston, DM. (1989) The product of the retinoblastoma susceptibility gene has properties of a cell cycle regulatory element. Cell 58: 1085-1095.

Draper GJ, Sanders BM, Kingston JE. (1986) Second primary neoplasms in patients with retinoblastoma. Br J Cancer 53:661-671.

Dryja TP, Rapaport JM, Joyce JM, Peterson RA. (1986) Molecular detection of deletions involving band q14 of chromosome 13 in retinoblastomas. Proc Natl Acad Sci USA 83:7391-7394.

Dunn JM, Phillip[s RA, Becker AJ, Gallie BL. (1988) Identification of germline and somatic mutations affecting the retinoblastoma tumors. Science 241:1797-1800.

Dyson N, Howley, PM, Munger, K, Harlow, E. (1989) The human papilloma virus-16E7 oncoprotein is able to bind to the retinoblastoma gene produce. Science 243:934-937.

Francke U. (1976) Retinoblastoma and choromosome 13. Berth Defects 12:131-137.

Friend SH, Bernards R, Rogelj S, Weinberg RA, Rapaport JM, Albert DM, Dryja TP. (1986) A human DNA segment with properties of the gene that predisposes to retinoblastoma and osteosarcoma. Nature 323:643-646.

Friend SH, Horowitz JM, Gerber MR, Wang X-F, Bogenmann E, Li FP, Weinberg RA. (1987) Deletions of a DNA sequence in retinoblastomas and mesenchymal tumors: organizations of the sequence and its encoded protein. Proc Natl Acad Sci USA 84:9059-9063.

Fung YKT, Murphree AL, T'Ang A, Qian J, Hinrichs SH, Benedict WF. (1987) Structural evidence for the authenticity of the human retinoblastoma gene. Science 236:1657-1661.

Harbour JW, Lai S-H, Whang-Peng J, Gazdar AF, Minna JD, Kaye FJ. (1988) Abnormalities in structure and expression of the human retinoblastoma gene in SCLC. Science 241:353-357.

Hensel DH, Hsieh, CLL, Gazdar, AF, Johnson, BE, Sakaguchi AY, Naylor SL, Lee WH and Lee EYHP. Altered structure and expression of the human retinoblastoma gene in small cell lung cancer. (submitted).

Hong FD, Huang HJS, To H, Young LJS, Oro R, Bookstein R, Lee EYHP, Lee WH. (1989) Structure of the human retinoblastoma gene. Proc Natl Acad Sci USA 86:5502-5506.

Horowitz JM, Yandell DW, Park SH, Canning S, Whyte P, Buchkovich K, Harlow E, Weinberg RA, Dryja TP. (1989) Point mutational inactivation of the retinoblastoma antioncogene. Science 243:937-940.

Huang HJS, Yee JK, Shew JY, Chen PL, Bookstein R, Friedmann T, Lee EYHP, Lee WH. (1988) Suppression of the neoplastic phenotype by replacement of the retinoblastoma gene product in human cancer cells. Science 242:1563-1566.

Huang S, Wang NP, Tseng B, Lee WH, Lee EYHP. Two distinct and frequently mutated regions of retinoblastoma protein are required for binding to SV 40 T antigen. (submitted)

Knudson AG. (1971) Mutation and cancer: statistical study of retinoblastoma. Proc Natl Acad Sci USA 68:820-823.

Lalande M, Dryja TP, Schreck RR, Shipley J, Flint A, Latt SA. (1984) Isolation of human chromosome 13-specific DNA sequences cloned from flow sorted chromosomes and potentially linked to the retinoblastoma locus. Cancer Genet Cytogenet 13:283-295.

Lee WH, Bookstein R, Hong F, Young LJ, Shew JY, Lee EYHP. (1987a). Human retinoblastoma susceptibility gene: cloning, identification, and sequence. Science 235:1394-1399.

Lee WH, Shew JY, Hong F, Sery T, Donoso LA, Young LJ, Bookstein R, Lee EYHP. (1987b) The retinoblastoma susceptibility gene product is a nuclear phosphoprotein associated with DNA binding activity. Nature 329:642-645.

Lee EYHP, To H, Shew, JY, Scully P, Lee WH. (1988a) Inactivation of the retinoblastoma susceptibility gene in human breast neoplasia. Science 238, 185-188.

Lee EYHP, Bookstein R, Young LJ, Lin CJ, Rosenfeld MG, Lee WH. (1988b) Molecular mechanism of retinoblastoma gene inactivation in retinoblastoma cell line Y79. Proc. Natl. Acad. Sci. USA 85, 6017-6021.

Lee WH, Bookstein R, Lee, EYHP. (1989a) Molecular biology of the human retinoblastoma gene. In "Tumor Suppressor Gene", G. Klein ed. New York, Marcel Dekker Inc.

Lee EYHP, Bookstein, R, Lee, WH. (1989b) Role of the retinoblastoma gene in the oncogenesis of human breast carcinoma. In "Breast Cancer: Cellular and Molecular Biology", Vol II. M.E. Lippman ed. Norwell, Kluwer Inc.

Ludlow JW, Decaprio JA, Huang CM, Lee WH, Paucha E, Livingston DM. (1989) SV40 large T antigen binds preferentially to an underphosphorylated member of the retinoblastoma susceptibility gene product family. Cell 56:57-65.

Matsunaga E. (1978) Hereditary retinoblastoma: delayed mutation or host resistance. Am J Hum Genet 30:406-424.

Mendoza AE, Shew J-Y, Lee EY-HP, Bookstein R, Lee W-H (1988) A case of synovial sarcoma with abnormal expression of the human retinoblastoma susceptibility gene. Hum Pathol 19:487-489.

McGee TL, Yandell DW, Dryja TP (1989) Structure and partial genomic sequence of the human retinoblastoma susceptibility gene. Gene 80:119-128.

Rappolee EA, Mark D, Banda MJ and Werb Z. (1988) Wound Macrophages express TGF-a and other growth factors in vivo: analysis by mRNA phenotyping. Science 241, 708-712.

Shew JY, Ling N, Yang X, Fodstad O, Lee WH. (1989) Antibodies detecting abnormalities of the retinoblastoma susceptibility gene product ($pp110^{RB}$) in osteosarcoma and synovial sarcomas. Oncogene Res 4:205-214.

Shew JY, Lin BTY, Chen PL, Tseng BY, Yang-Feng TL, Lee WH. C-terminal truncation of the retinoblastoma gene product leads to functional inactivation. (a, in press)

Shew JY, Chen PL, Bookstein R, Lee EYHP, Lee WH. Deletion of a splice donor site ablates expression of the following exon and produces an unphosphorylated RB protein unable to bind SV-40 T antigen. (b, in press)

Shields JA. (1983) "Diagnosis and Management of Intraocular Tumor." St. Louis: CV Mosby, pp 437-439.

T'Ang A, Varley JM, Chakraborty S, Murphree AL, Fung YKT. (1988) Structural rearrangement of the retinoblastoma gene in human breast carcinoma. Science 242:263-266.

T'Ang A, Wu K-J, Hashimoto T, Liu W-Y, Takahashi R, Shi X-H, Mihara K, Zhang F-H, Chen YY, Du C, Qian J, Lin Y-G, Murphree Al, Qiu W-R, Thompson T, Benedict WF, Fung Y-KT. (1989) Genomic organization of the human retinoblastpoma gene. Oncogene 4:401-407.

Toguchida J, Ishizaki K, Sasaki MS, Ikenaga M, Sugimoto M, Kotoura Y, Yamamuro T. (1988) Chromosomal reorganization for the expression of recessive mutation of retinoblastoma susceptibility gene in the development of osteosarcoma. Cancer Res. 48, 3939-3943.

Vogel F. (1979) Genetics of retinoblastoma. Hum Genet 52:1-54.

Varley JM, Armour J, Swallow JE, Jeffreys, AJ, Ponder BAJ, T'Ang A, Fung YKT, Brammar WJ, Walker RA. (1989) The retinoblastoma gene is frequently altered leading to loss of expression in primary breast tumors. Oncogene 4:725-729.

Wang NP, Chen PL, Huang S, Lee WH, Lee EYHP. DNA binding activity is intrinsic to the retinoblastoma gene product and is mapped to the C-terminal region of the protein. (submitted)

Whyte P, Buchkovich KJ Horowitz JM, Friend SH, Raybuck M, Weinberg RA, Harlow E. (1988) Association between an oncogene and an anti-oncogene:the adenovirus E1A proteins bind to the retinoblastoma gene product. Nature 334:124-129.

Whyte P, Williamson NM, Harlow E. (1989) Cellular targets for transformation by the adenovirus E1A proteins. Cell 56:67-75.

Wiggs J, Nordenskjold M, Yandell D, Rapaport j, Grondin V, Janson M, Werelius B, Petersen R, Craft A, Riedel K, Liberfarb R, Walton D, Wilson W, dryja TP. (1988) Prediction of the risk of hereditary retinoblastoma gene. N Eng J Med 318:151-157.

Yokota J, Akiyama T, Fung YKT, Benedict WF, Namba Y, Hanaoka M, Wada M, Terasaki T, Shimosato Y, Sugimura T, Terada M. (1988) Altered expression of the retinoblastoma (RB) gene in small-cell carcinoma of the lung. Oncogene 3:471-475.

Yunis JJ, Ramsay N. (1978) Retinoblastoma and subband deletion of chromosome 13. Am J Dis Child 132:161-163.

Index